WHAT EVERY ENGINEER SHOULD KNOW ABOUT CERAMICS

WHAT EVERY ENGINEER SHOULD KNOW
A Series

Editor

William H. Middendorf

*Department of Electrical and Computer Engineering
University of Cincinnati
Cincinnati, Ohio*

Vol. 1 What Every Engineer Should Know About Patents, *William G. Konold, Bruce Tittel, Donald F. Frei, and David S. Stallard*

Vol. 2 What Every Engineer Should Know About Product Liability, *James F. Thorpe and William H. Middendorf*

Vol. 3 What Every Engineer Should Know About Microcomputers: Hardware/Software Design: A Step-by-Step Example, *William S. Bennett and Carl F. Evert, Jr.*

Vol. 4 What Every Engineer Should Know About Economic Decision Analysis, *Dean S. Shupe*

Vol. 5 What Every Engineer Should Know About Human Resources Management, *Desmond D. Martin and Richard L. Shell*

Vol. 6 What Every Engineer Should Know About Manufacturing Cost Estimating, *Eric M. Malstrom*

Vol. 7 What Every Engineer Should Know About Inventing, *William H. Middendorf*

Vol. 8 What Every Engineer Should Know About Technology Transfer and Innovation, *Louis N. Mogavero and Robert S. Shane*

Vol. 9 What Every Engineer Should Know About Project Management, *Arnold M. Ruskin and W. Eugene Estes*

Vol. 10 What Every Engineer Should Know About Computer-Aided Design and Computer-Aided Manufacturing: The CAD/CAM Revolution, *John K. Krouse*

Vol. 11 What Every Engineer Should Know About Robots, *Maurice I. Zeldman*

Vol. 12 What Every Engineer Should Know About Microcomputer Systems Design and Debugging, *Bill Wray and Bill Crawford*

Vol. 13 What Every Engineer Should Know About Engineering Information Resources, *Margaret T. Schenk and James K. Webster*

Vol. 14 What Every Engineer Should Know About Microcomputer Program Design, *Keith R. Wehmeyer*

Vol. 15 What Every Engineer Should Know About Computer Modeling and Simulation, *Don M. Ingels*

Vol. 16 What Every Engineer Should Know About Engineering Workstations, *Justin E. Harlow III*

Vol. 17 What Every Engineer Should Know About Practical CAD/CAM Applications, *John Stark*

Vol. 18 What Every Engineer Should Know About Threaded Fasteners: Materials and Design, *Alexander Blake*

Vol. 19 What Every Engineer Should Know About Data Communications, *Carl Stephen Clifton*

Vol. 20 What Every Engineer Should Know About Material and Component Failure, Failure Analysis, and Litigation, *Lawrence E. Murr*

Vol. 21 What Every Engineer Should Know About Corrosion, *Philip Schweitzer*

Vol. 22 What Every Engineer Should Know About Lasers, *D. C. Winburn*

Vol. 23 What Every Engineer Should Know About Finite Element Analysis, *edited by John R. Brauer*

Vol. 24 What Every Engineer Should Know About Patents, Second Edition, *William G. Konold, Bruce Tittel, Donald F. Frei, and David S. Stallard*

Vol. 25 What Every Engineer Should Know About Electronic Communications Systems, *L. R. McKay*

Vol. 26 What Every Engineer Should Know About Quality Control, *Thomas Pyzdek*

Vol. 27 What Every Engineer Should Know About Microcomputers: Hardware/Software Design: A Step-by-Step Example. Second Edition, Revised and Expanded, *William S. Bennett, Carl F. Evert, and Leslie C. Lander*

Vol. 28 What Every Engineer Should Know About Ceramics, *Solomon Musikant*

Vol. 29 What Every Engineer Should Know About Developing Plastics Products, *Bruce C. Wendle*

Additional volumes in preparation

WHAT EVERY ENGINEER SHOULD KNOW ABOUT
CERAMICS

Solomon Musikant
*TransCon Technologies, Inc.
Paoli, Pennsylvania*

Marcel Dekker, Inc.   New York • Basel • Hong Kong

Library of Congress Cataloging--in--Publication Data

Musikant, Solomon.
 What every engineer should know about ceramics/Solomon Musikant.
 p. cm. -- -- (What every engineer should know; vol. 28)
 Includes bibliographical references and index.
 ISBN 0-8247-8498-7 (alk. paper)
 1. Ceramic materials. I. Title. II. Series: What every engineer should know; v. 28.
 TA455.C43M87   1990
 620.1'4-- --dc20   91-13916
                                                                    CIP

This book is printed on acid-free paper.

Copyright © 1991 by MARCEL DEKKER, INC. All Rights Reserved

Neither this book nor any part may be reproduced or transmitted in any form or by any means, electronic or mechanical, including photocopying, microfilming, and recording, or by any information storage and retrieval system, without permission in writing from the publisher.

MARCEL DEKKER, INC.
270 Madison Avenue, New York, New York 10016

Current printing (last digit):
10 9 8 7 6 5 4 3 2 1

PRINTED IN THE UNITED STATES OF AMERICA

*This book is dedicated to that human
who made the first ceramic molded figurine
24,000 years ago*

Preface

The purpose of this volume is to provide an overview of the rapidly advancing class of materials known as ceramics. Amazingly, human-made ceramic articles 24,000 years old are known, yet the technology of ceramics is a rapidly developing applied science in today's world. In fact there is keen competition among the leading industrial nations to exploit this science to the fullest.

The modern engineer or scientist encounters new developments daily. It is virtually impossible to be fully knowledgeable in even the limited areas that impact on one's individual field of endeavor. However, any technologist who has to deal with materials needs to be at least conversant with what is going on in the discipline of engineering ceramics.

Revolutions are taking place in which advanced ceramics play critical roles. A few such areas include the Space Shuttle, superconductivity, nuclear reactors, advanced gas turbines and reciprocating engines for energy conservation, integrated circuits, the laser, advanced optics, fiber optics, and biomedical applications. Each of these application areas represents truly amazing changes in the modern world.

PREFACE

This volume reviews the evolution of the ceramic technology and the early influences leading to today's worldwide interest in this arena. Although not intended to be a design manual, property tabulations and discussions of the major issues leading to successful applications are provided. The subjects covered include traditional ceramics, the new ceramics, ceramic processing, structural design considerations, the concept of fracture toughness (a central issue in ceramics), joining of ceramics, nondestructive testing and its importance, ceramic cutting tools and their implications, superconductive ceramics, advanced automotive ceramics, and carbon-carbon composites.

The structure of the ceramic crystal is complex and leads to many different forms. When one considers the number of atom types and arrangements that can be synthesized into ceramic bodies, it is easy to appreciate the fact that there is an infinite number of possibilities for the properties of such structures. That is why the developments in ceramics are leading to astounding discoveries and accomplishments. In the future, more and more variables will be discovered, studied, and applied, making possible even more revolutionary and useful applications.

Solomon Musikant

Contents

Preface   *v*

1	Ceramics Fundamentals	1
2	Traditional Engineering Ceramics	9
3	The New Ceramics	31
4	Processing of Ceramics	53
5	Structural Design Considerations	79
6	Fracture Toughness	99
7	Joining of Ceramics	123
8	Nondestructive Testing	129
9	Ceramic Cutting Tools	137
10	Space Shuttle Insulation Tiles	147
11	Superconductive Ceramics	153

12	Electronic Ceramics	**161**
13	Advanced Automotive Ceramics	**165**
14	Carbon-Carbon Composites	**185**

References *195*

Index *199*

WHAT EVERY ENGINEER SHOULD KNOW ABOUT
CERAMICS

1
Ceramics Fundamentals

Ceramics is commonly defined* as the art that deals with the design and fabrication of objects made from fired clay. An ancient piece of utilitarian earthenware fabricated about 1200 B.C. is shown in Figure 1.1; Figure 1.2 shows modern porcelain figurines of great delicacy and beauty. These products are about 3200 years apart in time but not so far apart in the principles of their fabrication or in the degree to which they are appreciated. All types of earthenware, stoneware, and porcelain are included in the term "ceramics." *Porcelain* refers to wares that are fired at high temperatures and are translucent, while *stoneware* and *earthenware,* such as terra-cotta, are fired at successively lower temperatures and are opaque.

More specifically, *ceramic* is defined as any of various hard, brittle, heat- and corrosion-resistant materials made by firing clay or other minerals and consisting of one or more metals in combination with one or more nonmetals, usually including oxygen. This definition must include not only pottery, but refractories (i.e., high-

---

*American Heritage Dictionary (Boston: Houghton Mifflin Co., 1978).

Figure 1.1 Ancient earthenware jug, approximately 1200 BC, Israel. (Photograph by S. Musikant.)

temperature resisting), structural clay products, ceramic coatings, abrasives, glass, glass-ceramics, and certain types of electronic compounds. In fact, a material such as silicon carbide, in which both the silicon and the carbon atoms have some metallic properties, is still considered a ceramic, although here there is some overlap with the definition of a semiconductor.

Glass is a special case. Although we have included glass in the definition of ceramic, the major distinction is that glass is an amorphous material, whereas ceramics are primarily crystalline in nature.

CERAMICS FUNDAMENTALS

Figure 1.2 Porcelains modeled by Arno Malinowski for the Royal Copenhagen Porcelain Manufactory. Figures, left to right: Asia-Europe, Africa, America, Australia. These important porcelains mark a departure from the forms of earlier porcelain ware. (Courtesy of the Cooper-Hewitt Museum of Design, Smithsonian Institution.) From "Ceramic Masterpieces" W. David Kingery and Pamela B. Vandiver, Free Press 1986.

However, most common ceramics have glassy phases incorporated in their microstructures. Glass-ceramics are another class of materials in which very small ceramic crystals are incorporated in a matrix of glass. The tiny crystallites impart desirable mechanical and thermostructural properties. Pyroceram is an example of a glass-ceramic.

Magnesium oxide (MgO) is an example of a simple oxide ceramic that is used as a high-temperature insulation in the form of blocks with some structural strength (i.e., a refractory) and that has a melting point of 2800 °C (5072 °F). It is a commonly used material for high-temperature industrial furnaces.

EVOLUTION OF THE CERAMIC TECHNOLOGY

The earliest known clay figures have been dated at about 22,000 B.C. Figure 1.3 illustrates some of the known history of ceramic art. As can be seen, every age contributed to the body of ceramic know-

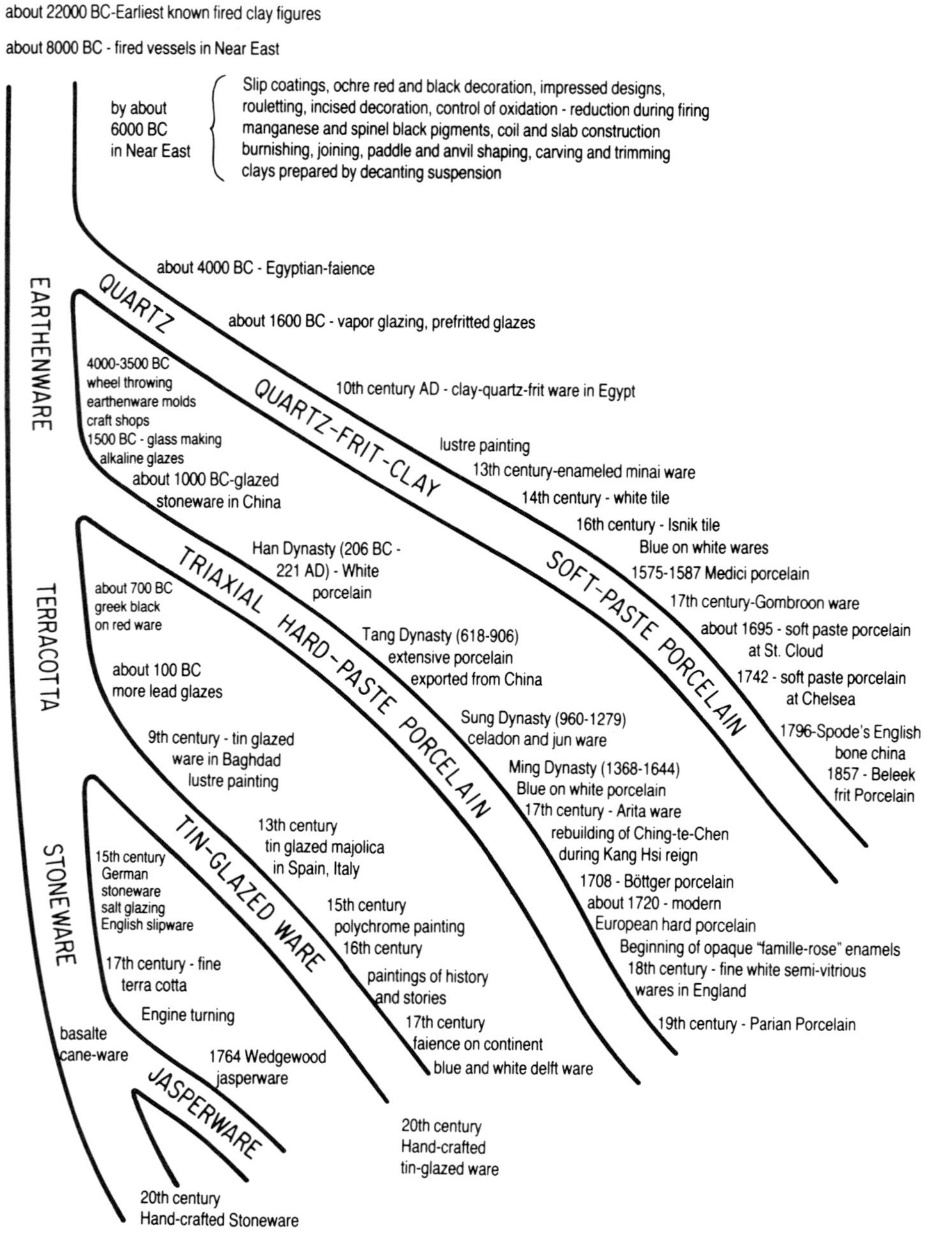

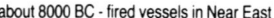

Figure 1.3 The flow of ceramic history illustrates the mainstreams of earthenware, terra-cotta, and stoneware; to "triaxial" hard-paste porcelain; of quartz-based bodies; and of tin-glazed ware. Some important shaping and decorative techniques are illustrated, but the diagram is far from complete. From "Ceramic Masterpieces" W. David Kingery and Pamela B. Vandiver, Free Press 1986.

CERAMICS FUNDAMENTALS

ledge. This storehouse of information is culminating in today's rapid technological developments in the field of ceramic materials.

Like many human endeavors, ceramics is an art whose beginnings are shrouded in the misty past, probably before invention of the written word. This is not a unique situation because it is easy to say the same thing about many of our "new" and revolutionary technologies. Examples include weaving, agriculture, metallurgy, the use of fire, mathematics, astronomy, weather forecasting, navigation, the sail (aerodynamics), mechanics (the lever, the wedge, the wheel, and the inclined plane), and as already stated, ceramics. The word itself comes from the Greek *keramos* pottery.

The industrial revolution was made possible by advanced furnaces and heat engines, and ceramic materials were essential for thermal insulation of the various types of furnaces and engines. In England, Wedgewood developed mass production techniques for ceramics that put useful and beautiful ceramic dishes within the reach of many. During the nineteenth and twentieth centuries the scientific understanding and manufacturing arts for the production of ceramic articles reached a high degree of sophistication. There became available a wide variety of new types of building materials with superior durability, strength, and other properties. These materials included brick, tile piping for drainage systems and roofing, sanitary ware, which was a primary factor in the development of public sanitation, and refractory (high-temperature) insulation materials, which served as furnace linings for the glass, steel, and other industries that depended on high-temperature processes. Rock wool is an early example of a ceramic fiber used to insulate buildings and appliances.

The raw materials for most, if not all, of these products came from mines and quarries, and these raw ingredients were prepared for the thermal processes needed to convert them to useful articles by crushing, washing, sieving, and mixing appropriate formulations. Usually, these naturally found materials were not pure, and the formulas had to take into account the small fractions of naturally occurring and often variable impurities and minor fractions.

Dielectric (electrically insulating) materials were important as the electrical and electronic technologies matured during the present century. These ceramic electrically insulative materials were developed to specifications set up by electrical and electronic device designers. As higher and higher frequencies and voltages were employed, the demands on ceramic dielectrics became more stringent. In addition, new specifications for the magnetic and optical properties of ceramics were developed as part of the new electronic and electro-optical technology revolution, and new classes of ceramic materials were devised to satisfy these requirements. This class of materials exhibits a complex variety of properties that are leading continually to new applications and classes of useful devices. High-temperature superconductors are a recent example.

CURRENT TRENDS

As the goals for more efficient generation of power from mobile heat engines were generated by the aerospace and automotive technologies, the intrinsic capability of metals to operate in the ever-higher-temperature regimes was exhausted. Higher temperatures, of course, mean higher operating efficiencies, with a tremendous payoff in terms of reduced fuel consumption. Certain classes of ceramic materials do have the ability to function as highly stressed structural members at much higher temperatures than metals. Therefore, engine designers looked to this class of material to solve the problem of higher-temperature operation. An extremely intricate ceramic turbine wheel designed to operate at about 1200 °C (2192 °F) is shown in Figure 1.4. In addition, tool builders are using ceramic cutting bits for extremely high cutting rates, and ceramic coatings are used to protect metals from high-temperature oxidation effects and to inhibit wear and erosion, as, for example, oil well drill bits, which have to bore through rock formations.

As we all know from common experience, ceramics are brittle and are susceptible to mechanical stresses or those induced by thermal shock. We have all seen a glass tumbler fracture when boiling water is poured into it. The fracture occurs because the stresses induced by the thermal gradients in the structure exceed the strength of the material. If the glass were warmed gradually from room temperature to 212 °F, thermal gradients would not develop and the glass

Figure 1.4 High-performance ceramic turbine wheel developed in collaboration with United Turbine in Sweden, utilizing the ASEA-patented glass encapsulation technique for near-net shape. From ASAE pamphlet Aϕ 20-104 E.

would survive. Sharp variation in the temperature of the tumbler wall causes uneven thermal expansion that sets up stresses. Since the glass cannot stretch to accommodate the variations in thermal expansion, high stresses and fracture occur. However, if the ceramic has a very small thermal expansion coefficient, the ceramic can survive even extreme thermal shock. One of the current thrusts in ceramic development is the radical improvement in the ability of ceramics to withstand such stresses for long periods of time, at high temperatures, and in adverse environments.

In the electrical world, invention of the transistor and laser has unleashed a flood of new devices all of which are limited in their ultimate performance by the materials available. It is a fact of nature that ceramics have the inherent capabilities to meet many of these needs, and advanced ceramics are finding their way into microelectronics, laser systems, communication devices and networks, magnetic components, and a variety of sensor applications.

Recent market studies have concluded that by the end of the current century there will be an order-of-magnitude greater use of ceramics of all sorts than now exists. There is already an army of scientists and engineers at work, especially in the United States, England, West Germany, and Japan, to solve the technical challenges imposed by this new materials revolution. Similar work, although not as widely publicized, is probably ongoing in the Eastern European nations.

The greatest immediate opportunity for high-volume application is in automotive engine applications. As the ceramic science advances and as the economies of scale become evident from the experience of the gigantic world automotive market, ceramics will find their way into many other types of markets. Sooner or later, every engineer will need to consider the effect that these advanced ceramics can have on the performance and the cost of a particular area of enterprise. An even more profound question will be: Can this new materials revolution spark a revolution in my own technology?

2
Traditional Engineering Ceramics

To appreciate the significance of the current technology of the new engineering ceramics, or *fine ceramics* as the Japanese have designated these advanced materials, it is useful to develop an appreciation of how the use of ceramics evolved and what the driving forces were that propelled this evolution. To generalize, the availability of materials with appropriate properties is the critical factor that has permitted human beings to create objects, machines, and structures of usefulness through all the millenia of humankind's advance from its inception. At first the natural materials we found around us were adequate for our needs, but as the knowledge, skills, and experience developed, these natural materials were not sufficient to meet the requirements of our increasingly ambitious projects.

So we are aware that the arts of rope making, weaving, boat building, shelter construction, agriculture, and cooking developed even though limited by the properties of natural materials. But even before the advent of written history, although recent in terms of human evolution, we discovered how to synthesize improved materials by the use of fire. The inventions of furnaces that made high temperatures available enabled truly revolutionary advances in metallurgy and in glasses and ceramics at an early date in human history. Metallurgy is a technology that preceded the development of ceramic technology and from which ceramics borrowed much.

STONE AS A CERAMIC

Stones and rocks are ceramics that were formed without the intervention of human intelligence. The huge cathedrals of medieval Europe were built from such materials to gigantic proportions as the architects and engineers learned how to use the natural stone. Some of the cathedrals collapsed during construction, teaching (the hard way) the limits of design and materials. The invention of the flying buttress was used to great advantage in the Notre Dame Cathedral, with its twin 204-foot-high towers (Figure 2.1). The realization that the strength of even the same type of stone was variable led to a comparison of the strength of the stone from various quarries. This was a critical milestone in the success of cathedral building technology. These endeavors culminated in the great church of Saint Peter in Rome, which was 619 feet long, 449 feet wide along the transepts, and a magnificent 470 feet high to the top of the cross. The foundation stone was laid on April 18, 1506, and the structure was completed November 18, 1626, one hundred and twenty years later—six human generations.

Granite is the great construction material. Composed of feldspar, quartz, and mica, it is among the most durable and strongest of the rocks, with a typical compressive strength of 19,000 psi and a modulus of elasticity of 7.3×10^6 psi. However, as with all stone as well as ceramic, the modulus of rupture (the bending strength) is far lower than the compressive strength—typically, 1800 psi. Table 2.1 lists typical mechanical properties of certain natural rocks as well as bricks. Due to its great compressive strength, a material such as granite has to be used in compression as much as possible, while minimizing tensile and bending stresses. The arch is an invention that allows a span to carry load while keeping the stone in compression. Figure 2.2 provides examples of both ancient and modern arch design.

TRADITIONAL ENGINEERING CERAMICS 11

Figure 2.1 Notre Dame Cathedral, Paris, France. (Photograph by S. Musikant.)

These basic ideas are true for ceramic as well as for successful stone designs. However, modern structural ceramics are far stronger than the granites, and a large part of the thrust of today's research and development is to provide ceramics with high strength and especially fracture toughness (i.e., resistance to propagation of preexisting flaws such as cracks; see Chapter 6).

THE BEGINNINGS OF CERAMIC TECHNOLOGY

In this book we discuss primarily crystalline ceramics. There is a host of knowledge relative to amorphous glass with compositions similar to the ceramic materials under consideration here, but a compre-

Table 2.1 Typical Mechanical Properties of Stone and Brick[a]

	Specific gravity	Compressive strength (psi)	Modulus of elasticity (psi)	Absorption of water (parts by weight)	Coefficient of expansion (per °F)	Modulus of rupture (psi)
Granite	2.67	19,400	7,300,000	1/750	0.0000040	1,850
Limestone	2.53	9,500	8,460,000	1/38	0.0000045	1,400
Limestone, oolithic	2.48	6,700	7,000,000	1/23	0.0000045	
Marble	2.72	12,700	8,000,000	1/300	0.0000045	1,400
Sandstone	2.22	9,300	3,000,000	1/24	0.0000055	1,400
Trap	2.96	20,000	12,000,000			
Slate	2.77	14,000	14,000,000			
Brick						
Common	2.00	4,000	2,000,000	1/3		
Hard-burned	2.10	8,000	4,000,000	1/6		
Paving	2.42	10,000	7,000,000	1/100		
Sand-lime	1.85	3,500	1,000,000	1/10		
Brick masonry						
In lime mortar		0.14 ×2 compressive strength of brick				
In cement mortar		0.23 × compressive strength of brick				

[a]Other strength functions: Shearing strength of brick and stone is from 10 to 20% of the compressive strength; tensile strength is 4% of compressive strength; modulus of rupture is 15% of compressive strength. Poisson's ratio is 1/4. (From *Mark's Mechanical Engineer's Handbook*, McGraw Hill (1958).

hensive discussion of glass is outside the scope of this volume*. However, it must be recognized that many engineering ceramic materials as well as natural rocks are actually agglomerates of microscopic crystals held together by an intragranular glassy phase. Figure 2.3 shows the microstructure of one type of polycrystalline ceramic found in an eighteenth-century porcelain. The microstructure shows crystals of quartz and mullite in a glassy matrix. Figure 2.4, on the other hand, shows the microstructure of a nearly pure alumina (Al_2O_3) that possesses a fine-grained structure. There are many different types of microstructure in the full realm of ceramic bodies.

Traditional engineering ceramics are divided into several major classes: refractories for application in high-temperature furnaces

---

*See Solomon Musikant, Glass, in *Encyclopedia of Physical Science and Technology* (New York: Academic Press, 1986).

Figure 2.2 (a) Supporting arches, King Solomon's Stables. The circle was the basis of the arch in ancient middle eastern structures. (b) Modern arch, Philadelphia Academy of Fine Arts. (Photograph by S. Musikant.)

MIT ceramics
10 μm

(a)

Silica-rich glass (Quartz has dissolved)
Partially dissolved quartz grain surrounded by silica-rich glass
Larger mullite needles and glass form from feldspar grain
Fissures form around quartz during cooling
Fine mullite crystals in glassy matrix form from clay

(b)

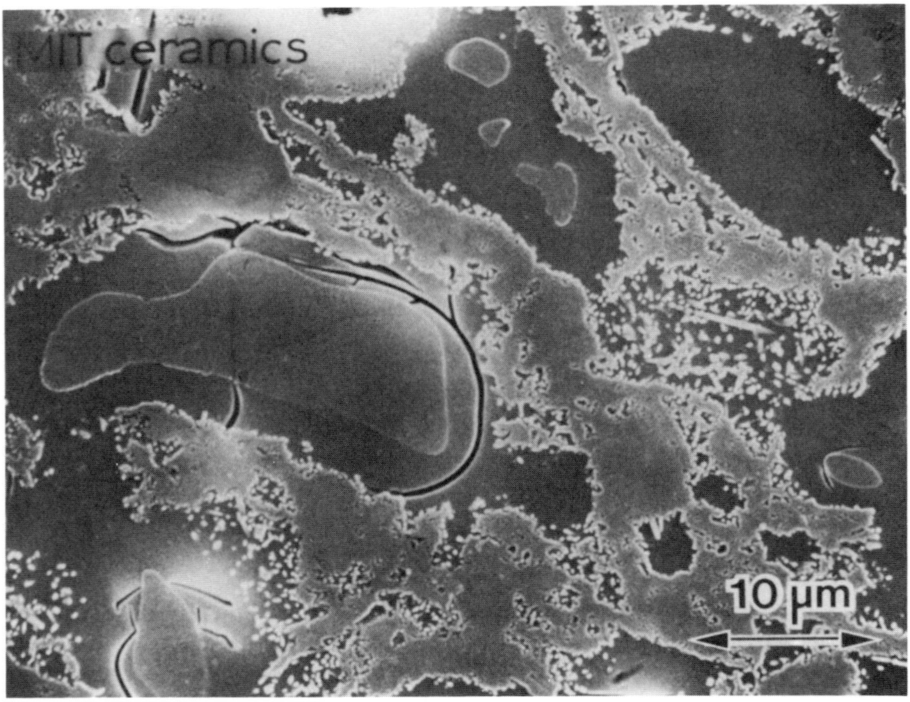

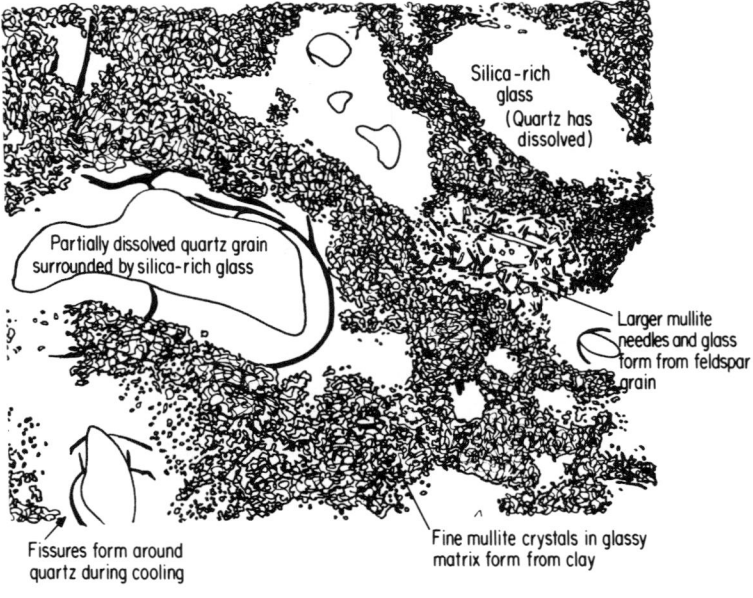

TRADITIONAL ENGINEERING CERAMICS

as structure and as thermal insulation; whiteware, which includes porcelain, sanitary ware, electrical insulation, and dishes; cement; the structural clay materials, which include bricks, tiles, and sewer pipe; and abrasives, such as silicon carbide and aluminum oxide. In addition, there are many and diverse special application areas.

In the case of the refractories, not only are the thermal and structural properties of importance, but the chemical resistance of the ceramic to the substances and gases that come in contact with the ceramic are of vital significance. For example, the ceramic liners for glass tanks that are in contact with the extremely corrosive molten glass must last for a year or more with limited dissolution to be attractive economically. The same can be said for the refractories composing the crown (or ceiling) of the glass furnace which must withstand the corrosive combustion gases and volatile materials coming from the batching ingredients.

Whiteware is generally made from a mixture of clay, feldspar, and flint. Clays consist predominantly of aluminosilicates with water of hydration. They are fine grained and the clay particles tend to slip easily across each other under shear stress. Feldspar is also an aluminosilicate with other elements, such as potassium, sodium, or calcium, included in its structure. When heated, these metallic components form a flux that aids in formation of the glassy phase, which binds the ceramic particles together in the fired ceramic body. Feldspar is a common mineral. Flint is primarily silica (SiO_2) with traces of iron, calcium, carbonaceous substances, and sometimes gypsum (hydrated calcium sulfate with various impurities).

These starting minerals are all abundant and cheap and can be formulated and processed in various ways to yield an almost infinite

---

Figure 2.3 The microstructure of eighteenth-century K'ang Hsi Chinese porcelain. Large quartz particles are mostly dissolved in the alkali-silicate glass. Tiny crystals of mullite derived from the clay material, and also from the feldspar, are immersed in the continuous glass phase. The fissures adjacent to the quartz grains (exaggerated by the etchant used) result from differential contraction on cooling (1000×). The sketch is provided to help identify the constituent phases present. [From W. D. Kingery and P. B. Vandiver, *Ceramic Masterpieces* (New York: The Free Press, 1986), p. 22.]

Figure 2.4 Surface replica of an as-fired fine-grained 99% Al_2O_3 substrate surface. [From W. D. Kingery, H. K. Bowen, and D. R. Uhlman, *Introduction to Ceramics* (New York: John Wiley & Sons, 1976), p. 565.]

variety of whiteware. Table 2.2 lists a variety of whiteware bodies and the typical compositions for each. The cone number refers to the firing temperature: The higher the cone number, the higher the temperature to which the raw material is subjected. In the category of "other" ingredients, the bone ash provides calcium oxide and the talc, magnesia (MgO), to the starting composition. Each class of whiteware is tailored to provide the combination of properties and economy demanded by the application indicated. Sanitary ware is actually heated high enough to become vitrified (made glassy), to be completely nonabsorbent.

The manufacture of portland cement dates from the middle of the nineteenth century. Although the Romans knew of a similar

Table 2.2 Composition of Triaxial Whiteware Compositions

Body type[a]	China clay	Ball clay	Feldspar	Flint	Other
Cone 16 hard porcelain	40	10	25	25	—
Cone 14 electrical insulation ware	27	14	26	33	—
Cone 12 vitreous sanitary ware	30	20	34	18	—
Cone 12 electrical insulation	23	25	34	18	—
Cone 10 vitreous tile	26	30	32	12	—
Cone 9 semivitreous whiteware	23	30	25	21	—
Cone 10 bone china	25	—	15	22	38 bone ash
Cone 10 hotel china	31	10	22	35	$2 CaCO_3$
Cone 10 dental porcelain	5	—	95	—	—
Cone 9 electrical insulation	28	10	35	25	2 talc

[a] In order of decreasing firing temperature.

process, the art died out and was rediscovered in England in the latter part of the eighteenth century. Cement is made from limestone mixed with additions of silica (SiO_2), alumina (Al_2O_3), and ferric oxide (Fe_2O_3) and heated (burned) to a high enough temperature to remove the water of hydration. The clinkers so formed are ground to a fine powder. When water is added to the dehydrated material, a relatively strong bond is formed as the water combines chemically with the ground residue of the firing process and drying or curing of the mass takes place. Concrete is the mixture of various particle sizes of stone embedded in cement to form a composite stronger and with better wear-resistant properties than those of the cement alone.

The structural clays are mainly the brick, tile, and sewer pipe products used in the construction industries. Brick is essentially a fired clay product. The quality of brick is associated with the clay composition and the firing temperature used in its production. The durability of brick in northern climates is a function of its strength, water absorption, and solubility. Obviously, water absorption leads to ice formation in the pores of the ceramic in the winter and spallation stresses due to the increase of volume of the absorbed material as it freezes. A brick for an exterior wall typically needs to have water absorption less than 10% and a compressive strength greater than 3500 psi to be durable.

Similarly, tiles are made to different specifications depending on application. Drain tile needs to have high porosity, while sewer pipe is designed to avoid leakage and is often glazed with a glassy, relatively impervious coating. Floor and roof tile must be hard and dense to possess the strength and durability required by the forces these products see in service, and in the case of roof tile, to be resistant to water penetration. The glaze is made from finely ground components suspended in water and applied to the surface of the clay product after an initial firing while it is still porous. The tile is then dried and refired to a temperature sufficiently high to fuse (melt) the applied coat to form a glassy finish.

Porcelain is the name given to thin-ware in which the glaze coating completely penetrates the body and, after the fusing process, causes the article to be translucent, thus imparting visual beauty.

OXIDES, NITRIDES, AND CARBIDES

Most of the discussion above dealt with ceramics made from multicomponent compositions. There is a wide variety of ceramics that are simple compounds. In this category fall the oxides, the nitrides, and the carbides, which are well known and widely used classes of ceramics. In general, these products are used for various special purposes and are more expensive than products made from the inexpensive raw materials obtained directly from quarries and mines. Most of the binary compounds mentioned here require specialized and sometimes involved processing to prepare them from the available raw materials.

The preparation of these materials in production quantities evolved at a later date than the traditional multicomponent ceramics as the industrial age's technical development led to more complex applications and greater insistence on very specialized materials. It is this class of materials that is now in the forefront of development and application as greater understanding is achieved of the properties, preparation techniques, and modes of application of these materials. More will be said about some of these in later chapters.

Table 2.3 Mechanical and Electrical Ceramics: Fired Parts

	Pyroceram 9606	Polycrystalline glass (glass-ceramic)			Cordierite	Forsterite
		Pyroceram 9608	Pyroceram 9658			
Physical properties						
Specific gravity	2.61	2.50	2.52		2.29-2.65	2.9
Ther. cond. (Btu/hr/ft²/°F/ft)	1.95	1.13	0.97		0.97-2.40	1.94-2.40
Coef. of ther. exp. (per °F)						
68-212°F	—	—	—		—	—
77-570°F	3.2×10^{-6}	$0.2\text{-}1.1 \times 10^{-6}$	6.2×10^{-6}		2.08×10^{-6}	4.72×10^{-6}
68-932°F	—	—	—		1.68×10^{-6}	5.40×10^{-6}
Thermal shock resistance	Good	Good	Good		Excellent	Moderate
Water absorption, (%)	0.00	0.00	0.00		0.02-3.2	0.00-0.02
Gas permeability	Gastight	Gastight	Gastight		—	Gastight
Spec. ht. (Btu/lb/°F)	0.233	0.190	0.20		—	—
Mechanical properties						
Mod. of elast. in ten. (psi)	17.3×10^6	12.5×10^6	9.3×10^6		7×10^6	—
Mod. of rupture (1000 psi)	20	12-15	15		6.8	19
Ten. str. (1000 psi)	—	—	10		3.0	10
Hardness (Knoop)	657[a]	593[a]	250		7 (Mohs)	7.5 (Mohs)
Impact str. (Charpy) (in.-lb)						
¾-in. dia.	—	—	—		7.4	7.5
½-in. dia.	—	—	—		4.4	2.4-4.0
Compr. str. (1000 psi)	—	—	50		52-95	80-85
Electrical properties						
Volume res. (Ω-cm)						
68-77°F	2×10^{16}	6.3×10^8	1×10^{14}		$>10^{14}$	$>10^{14}$
212°F	1.6×10^{13}	1.3×10^{11}	—		—	5.0×10^{13}

(continues)

Table 2.3 (continued)

	Pryoceram 9606	Polycrystalline glass (glass-ceramic) Pyroceram 9608	Pyroceram 9658	Cordierite	Forsterite
482 °F	10^{10}	1.25×10^8	—	—	—
572 °F	2×10^9	2×10^7	—	—	7.0×10^{11}
662 °F	6×10^8	6.31×10^6	—	—	—
932 °F	2×10^7	3.1×10^5	1×10^7	—	1.2×10^{10}
1292 °F	—	—	—	—	1.0×10^8
1652 °F	—	—	—	—	3.0×10^6
Dielectric str. (V/mil)	1000	—	1000	140-230	250
Dielectric constant					
100 cycles	5.62	7.13	—	—	6.3^d
1 Mc	5.58	6.78	—	4.02-6.23	6.2-6.5
100 Mc	—	—	—	—	6.1
10,000 Mc	5.45	6.55	5.92	—	5.8
Dissipation factor					
100 cycles	0.0016	0.020	—	—	0.0014^d
1 Mc	0.0015	0.0030	—	0.0010-0.009	0.0002-0.0004
100 Mc	—	—	—	—	0.0003
10,000 Mc	0.00033	0.0068	0.003	—	0.0010

Loss factor				
100 cycles	0.009	0.14	—	0.009[d]
1 Mc	0.008	0.02	0.0297-0.0579	0.001-0.002
100 Mc	—	—	—	0.002
10,000 Mc	0.002	0.045	—	0.0058
T_e value (°F)[b]	1400	815	1436	680 – >1832
Temp. coef. of capacitance chg.[c]	—	—	420	—
Heat Resistance				
Max. recommended svc. temp. (°F)	—	1800	—	1832
Uses	Developed for uniform electrical properties in missile radomes; suitable for high-temperature, high-frequency applications in electronics	General purpose; line of shock resistant cooking-serving ware; telescope mirror blanks	Improved machinability for precision shapes; vacuum seals and enclosures, thermal and electrical insulators	Aircraft firewall connectors, appliance coil supports and terminal blocks, automotive heater cores, hot-point insulators, brazing fixture parts, foundry parts, fuel burner tips, thermostat controls, catalytic converter substrate	Very low loss insulators, ceramic-to-metal seals (close tolerances obtainable by grinding)

[a] Knoop hardness at 100 g.
[b] T_e value is the temperature at which 1 cm^3 of the material has a resistance of 1 MΩ.
[c] 77-185 °F, parts per million.
[d] 60 cycles.

Table 2.4 Mechanical and Electrical Ceramics: Fired Parts

	Standard electrical	Refractory mullite	Steatite	Zircon	Alumina
Physical properties					
Specific gravity	2.37-2.53	3.0-3.3	2.5-2.92	3.43-3.86	3.4-3.9
Ther. cond. (Btu/hr/ft^2/°F/ft)	0.87-1.57	1.38-1.45	1.45-1.94	2.88-3.61	13.3-21.3
Coef. of ther. exp. (per °F)					
68°-212°F	2.00×10^{-6}	$2.7\text{-}3.0 \times 10^{-6}$	$3.33\text{-}3.99 \times 10^{-6}$	$1.31\text{-}1.84 \times 10^{-6}$	$3.0\text{-}3.3 \times 10^{-6}$
68°-932°F	2.70×10^{-6}	—	$4.52\text{-}5.50 \times 10^{-6}$	$2.09\text{-}2.16 \times 10^{-6}$	$3.9\text{-}4.4 \times 10^{-6}$
Thermal shock resistance	Fair	Good to excellent	Moderate	Good	Good
Water absorption (%)	0.0-0.1	0.00	0.0-1.0	0.0-9.0	0.000
Gas permeability	Gastight	Impervious	—	—	Impervious
Mechanical properties					
Mod. of elast. in tension (psi)	10×10^6	—	$13\text{-}16 \times 10^6$	21×10^6	$30\text{-}50 \times 10^6$
Tensile strength (1000 psi)	2.5-7.0	14-18	4.8-15	4.5-12	20-30
Hardness (Mohs)	7.0-7.5	7.5-9.0	7.5	8	9
Impact strength (Charpy) (in.-lb.)					
¾ in. dia.	8.2	—	10.5-14.0	8.9-11.4	—
½ in. dia.	—	—	3.8-5.0	5.50-5.64	6-8
Flexural strength (1000 psi)	5.4-12.0	—	11-20	18.5-22.0	30-60
Compressive strength (1000 psi)	49.1-70.0	100-150	66-90	60-100	140-400
Electrical properties					
Volume resistivity (Ω-cm)					
68-77°F	$10^{13}\text{-}10^{15}$	$10^{14}\text{-}10^{15}$	$>10^{14}$	$>10^{14}$	$>10^{14}$
212°F	1.2×10^{6a}	—	$0.21\text{-}>100 \times 10^{13}$	2.0×10^{13}	$10^{13} - >10^{14}$
570°F	5.0×10^6	—	$0.6\text{-}800 \times 10^8$	5.5×10^{11}	$10^{10} - >10^{14}$
930°F	4.0×10^{5b}	—	$0.32\text{-}300 \times 10^6$	5.5×10^8	$10^8\text{-}4.0 \times 10^{12}$
1290°F	—	10^6	$2.3\text{-}500 \times 10^6$	1.4×10^7	$10^6\text{-}3.0 \times 10^9$
1650°F	—	—	$7.0\text{-}680 \times 10^3$	8.2×10^5	$2 \times 10^5 - 7 \times 10^9$
Dielectric strength (V/mil)	55-300	300	145-280	60-290	250
Dielectric constant					
60 cycles	5.4-7.0	—	5.9-6.3	9.1	—

1 Mc	—	6.5-7.0	5.5-6.51	5.30-9.20	8.0-10.0
100 Mc	—	—	5.6-6.0	8.6	8.0-10.0
10,000 Mc	—	—	5.3-5.8	8.4	7.5-9.6
Dissipation factor					
60 cycles	0.0090-0.0112	—	0.0013-0.0150	0.0360	—
1 Mc	—	—	0.0011-0.0075	0.0007-0.0022	0.0001-0.0009
100 Mc	—	—	0.0009-0.0028	0.0012	—
10,000 Mc	—	—	0.0014-0.0054	0.0027	0.002-0.0040
Loss factor					
60 cycles	0.053-0.060	—	0.008-0.090	0.327	—
1 Mc	—	—	0.007-0.0252	0.0041-0.0135	0.0008-0.0090
100 Mc	—	—	0.005-0.016	0.010	—
10,000 Mc	—	—	0.008-0.030	0.023	0.0015-0.0864
T_e value (°F)	680-842	1292-1472	824-1544	1292-1598	1290->2000
Temp. coef. of capacitance chg.[a]	630	—	120	175	120
Heat resistance					
Max. recommended Svc. Temp. (°F)	1820	3000-3200	1832	2012	2500-3200
Uses	Low-voltage insulators, vitrified high-voltage insulators, wire supports, outlet boxes, lightning arrestors, suspension insulators, X-ray rods and tubes	High-temperature insulators, spark plugs, laboratory ware	Appliance housings, automotive electrical ballast, electric line insulators, tube sockets, electrical instrument spacers and feed-through bushings, fuel igniters, camera and projector rollers, thermostat controls	Aircraft firewall connector plugs and glow plugs, electronic tube sockets, coil forms, brackets, printed circuits and plates, pump valves, plungers and seats	Spark plugs, substrates, tube envelopes, radomes, bushings, terminals, pump plungers, valve seats, high-frequency insulations, water pump seals

[a] 392 °F.
[b] 752 °F.
[c] 77-185 °F, parts per million.

Table 2.5 Carbides and Nitrides: Fired or Sintered Parts

	Carbides and Nitrides						
	Silicon carbide				Boron carbide	Silicon nitride	
	Silicate bonded	Silicon nitride-bonded	Fine-grain reaction sintered (KT)	Sintered alpha SIC		Reaction bonded	Hot pressed
Physical properties							
Density (lb/in³)	0.093	0.095	0.107-0.108	0.113-0.115	0.087[a]	0.027-0.098	0.115
Porosity (%)	9-17	8-15	Negligible	Negligible	Negligible	15-76	Negligible
Ther. cond. (2200°F) (Btu/hr/ft²/°F/ft)	9	10	18.4	15.5	16	8	14-19
Coef. of ther. exp. (0-2550°F) (per °F)	2.4×10^{-6}	2.4×10^{-6}	2.04×10^{-6b}	2.23×10^{-6c}	1.73×10^{-6}	$1.4\text{-}1.8 \times 10^{-6}$	2×10^{-6}
Specific heat (0-2550°F) (Btu/lb/°F)	0.285	0.288	0.32[d]	0.31[d]	—	0.17	0.17
Max. service temperature (°F)							
Inert atmosphere	3200	3200	2550	3000	4100	2370	2370
Oxidizing atmosphere	2900	3000	2550	3000	1000	—	—
Mechanical properties							
Mod. of elast. in tension (77°F) (psi)	13.2×10^6	17×10^6	48.1×10^6	58.9×10^6	42×10^6	24×10^6	45×10^6
Tensile strength (77°F) (1000 psi)	1	3	—	—	22.5	—	—
Compressive strength (77°F) (1000 psi)	15	20	—	565	420	77-112	500
Modulus of rupture (77°F) (1000 psi)	2.2	5.5	55.6	66.6	45	30	—

[a] Boron carbide is available in densities ranging from 0.069 to 0.091 lb/in³.
[b] 0-1100°F; 2.57×10^{-6} above 1100°F.
[c] 0-1100°F; 2.94×10^{-6} above 1100°F.
[d] 0-2200°F.

Hot-pressed and sintered carbides

	Beryllium carbide (Be_2C)	Titanium carbide (TiC)	Columbian carbide (CbC)	Tantalum carbide (TaC)	Zirconium carbide (ZrC)
Ther. Cond. (68-795 °F) (Btu/hr/ft²/°F/ft)	12.1	9.9[a]	8.23[a]	12.8[a]	11.9[a]
Coef. of ther. exp. (77-1472 °F) (per °F)	5.8×10^{-6}	4.1×10^{-6}	—	4.6×10^{-6}	3.7×10^{-6}
Electrical resistivity (Ω-cm)[a]	1.1	1.05×10^{-4}	7.4×10^{-5}	2×10^{-5}	6.34×10^{-6}
Hardness (Mohs)	9+	8-9	9-10	9+	8-9
Compressive strength (1000 psi)[a]	105	109	—	—	238
Mod. of rupture (1000 psi)[a]	16	—	—	—	—
Ther. shock. res. (air quenched) (cycles)	4 at 2000-1470 °F	—	—	—	—
Fabrication methods	Hot pressing, steel die pressing and sintering, hydrostatic pressing and sintering	Hot pressing, steel die pressing and sintering	Hot pressing, steel die pressing and sintering	Hot pressing, steel die pressing and sintering	Hot pressing, steel die pressing and sintering

[a] Room temperature.

(continues)

Table 2.5 (continued)

	Carbides with metal binders			
	Titanium carbide (TiC)[a]	Tungsten-titanium carbide (WTiC$_2$)[b]	Tungsten carbide (WC)	Chromium carbide (Cr$_4$C, Cr$_7$C$_3$, Cr$_3$C$_2$)[c]
Density (lb/in.^{3})	0.20-0.26	0.38-0.47	0.40-0.55	0.25-0.29
Ther. cond. (68 °F) (Btu/hr/ft^2/°F/ft)	9.9-13.1	16.5-32.9	16-69	—
Coef. of ther. exp. (68-1200 F), (per °F)	4.3-7.5 × 10^{-6}	3.5-4.0 × 10^{-6}	2.5-3.9 × 10^{-6}	5.8-6.3 × 10^{-6d}
Electrical conductivity (% IACS)	1.34-6.0	4.3-5.8	4.3-11.9	2.58-2.78
Mod. of elast. in tension (psi)				
70 °F	42-65 × 10^6	65.5-80.6 × 10^6	61.6-94.8 × 10^6	—
1600-1800 °F	33-48 × 10^6	—	—	—
Tensile strength (1000 psi)[e]				
75 °F	26-134 (0-61)	118-145	130[f]	36-37 (0)
1500 °F	45-94 (0-2.7)	—	—	20-42 (0.2)
1800 °F	35-72 (0-2.4)	—	—	—
Hardness (Rockwell)	A73-A93.5	A90-A93	A85-A93	A86.5-A89
Impact strength (unnotched Charpy) (ft-lb)				
75 °F	1.5-16	5.3-8.9	—	—
1800 °F	2.5-16	—	—	—
Transverse rupture strength (1000 psi)	100-236	125-350	70-475	100-120
Stress-rupture strength (100 hr, 1800 °F) (1000 psi)	8-28	—	—	—
Compressive strength (1000 psi)	265-685	585-705	500-935	422-480

[a] Property range covers grades ranging from 17.5 to 90% TiC with different binder metal contents.
[b] Property range covers various grades of different carbide-metal proportions.
[c] The type of chromium carbide and the type of binder metal affects properties.
[d] 68-576 °F.
[e] Elongation (%) in parentheses.
[f] Typical of one grade.

TRADITIONAL ENGINEERING CERAMICS

Table 2.6 Refractory Ceramics and Cermets: Fired or Sintered Parts

	High-alumina ceramics[a]			Alumina cermets	
	Alumina content			Chromium-alumina	Molybdenum-chromium-alumina[a]
	85%	95%	99+%		
Physical properties				**Physical properties**	
Specific gravity	3.45	3.70	3.85	Density (lb/in.^{3}) — 0.21	0.22
Ther. cond. (200°F) (Btu/hr/ft^2/°F/ft)	8.5	12.1	14.5	Porosity, (%) — <0.25	<0.25
Coef. of ther. exp. (per °F)				Melting point (approx.) (°F) — 336	—
77-390°F	3.1 × 10^{-6}	3.7 × 10^{-6}	3.7 × 10^{-6}	Ther. cond. (Btu/hr/ft^2/°F/ft) — 29[a]	—
77-750°F	3.7 × 10^{-6}	4.0 × 10^{-6}	4.0 × 10^{-6}	Coef. of ther. exp. (per °F) — 4.7 × 10^{-6b}	5.2 × 10^{-6a}
77-1100°F	3.9 × 10^{-6}	4.3 × 10^{-6}	4.3 × 10^{-6}	Spec. ht. (calc.) (Btu/lb/°F) — 0.16	0.14
77-1470°F	4.1 × 10^{-6}	4.5 × 10^{-6}	4.5 × 10^{-6}	Poisson's ratio — 0.22	0.25-0.27
77-1830°F	4.3 × 10^{-6}	4.7 × 10^{-6}	4.7 × 10^{-6}	**Mechanical properties**	
Water absorption (%)	0.0	0.0	0.0	Mod. of elast. in tension (psi) — 41 × 10^6	37-39 × 10^6
Max. recommended svc. temp. (°F)	2460	2800	3000	Ult. ten. str. (1000 psi)	
Electrical properties				Room temp. — 21	—
Dielec. str. (V/mil)	210	210	220	800°F — 20.5	—
Dielec. const. (77°F, 1 Mc)	8.2	8.9	9.6-10.0	1200°F — 20	—
				1500°F — 19.7	—

(continues)

Table 2.6 (continued)

	High-alumina ceramics[a]			Alumina cermets	
	Alumina content			Chromium-alumina	Molybdenum-chromium-alumina[a]
	85%	95%	99+%		
Power Factor (77°F, 1 Mc)	0.0009	0.00035	0.00027	16.8	—
Loss Factor (77°F, 1 Mc)	0.007	0.003	0.003	11.7	—
T_e value (°F)	1560	1960	2012		
Mechanical properties					
1800°F				C37	C45-55
2000°F					
Hardness (Rockwell)					
Mod. of rupture (1000 psi)					
Room temp.				45	55[d]
1800°F				27	55
2100°F				18	29
2400°F				4.6	12
Mod. of elast. in ten. (psi)	32×10^4	40×10^4	50×10^4	110	240
Ten. str. (1000 psi)	20	30	39		
Flex. str. (1000 psi)	40	50	50-70		
Compr. str. (1000 psi)	250	300	400	17×10^4	15×10^4
Hardness				40	—
Mohs	9	9	9		
Knoop (1000-g load)	1450	1750	2000	21×10^4	26×10^4
Impact str. (Charpy) (in.-lb)	6.5	7.0	—		
Compr. str. (1000 psi)					
Mod. of rigidity (psi)					
Shear str. (1000 psi)					
Bulk modulus (psi)					

[a]The values given are not maximum values and are dependent on the minor components or fluxes used as well as a number of other factors.

[a]At 500°F.
[b]At 32-1832°F.
[c]At 68-1472°F.
[d]Addition of tungsten raises room temperature modulus of rupture to about 70,000 psi.

TRADITIONAL ENGINEERING CERAMICS

Refractory Oxides

	Alumina (Al$_2$O$_3$)	Beryllia (BeO)	Calcia (CaO)	Magnesia (MgO)	Thoria (ThO$_2$)	Zirconia (stabilized ZrO$_2$)	(vitreous SiO$_2$)
Melting point (°F)	3700	4620	4710	5070	6000	4710	—
Ther. cond. (st. spec. temp. and porosity) (Btu/hr/ft^2/°F/ft)	10.7 (200 °F, 0%)	95.2 (2190 °F, 5-10%)	4.12 (1830 °F, 9%)	1.47 (2190 °F, 22%)	0.0 (2190 °F, 17%)	0.53 (2190 °F, 28%)	0.80 (0.9%)
Coef. of ther. exp. (per °F)	43 × 10^{-7} (77-1830 °F)	52.8 × 10^{-7} (68-2550 °F)	75.5 × 10^{-7} (68-2190 °F)	77.8 × 10^{-7} (68-2550 °F)	52.8 × 10^{-7} (68-2550 °F)	30.6 × 10^{-7} (68-2190 °F)	2.8 × 10^{-7} (68-2280 °F)
Max. use temp. in oxidizing atm. (°F)	3540	4350	4350	4350	4890	4530	—
Hardness (Mohs)	9	9	4.5	6	7	7-8	—
Thermal shock resistance	Good	Excellent	Fair	Fair	Poor	Fair	Excellent
Stability in:							
Reducing atmosphere	Excellent	Excellent	Poor	Poor	Good	Good	Fair
Carbon	Excellent	Excellent	Poor	Good	Fair	Fair	Good
Acid slags	Excellent[b]	—	Poor	Poor	Poor	Good	Good
Basic slags	Excellent	Fair	Fair	Good	Good	Poor	—
Metals	Good	Good	Fair	Fair	Excellent	Good	—

[a]Depends on degree of stabilization.
[b]Except HF.

PROPERTIES

Such a variety of materials is encompassed in the discussion above that a complete properties listing would be a mammoth undertaking. In fact, no such compendium exists, although books have been prepared with various degrees of comprehensiveness. At this point, some properties of typical engineering ceramics will be presented. However, for design and analysis, readers are advised to seek data on the specific materials they are concerned with, as all the properties vary widely depending on the exact process and composition of manufacture. Tables 2.3 through 2.6 provide properties for a variety of traditional ceramic products as well as for the binary compounds discussed in this section.

3
The New Ceramics

There are many possible ceramics because there are many combinations of metallic and nonmetallic atoms that can combine to form ceramic compounds. In addition, several structural arrangements are usually possible for each combination of atoms. When one considers further that ceramics can be multielement and multiphase with a considerable number of different atoms entering into the structure, the possibilities for new materials becomes infinite. However, in practice it takes much in the way of both resources and time to perfect even a single ceramic. Therefore, only a relatively few materials have been or are being carried through to development completion. Constant reassessment of progress is made to determine which candidate material should get the available resources. Or in business management parlance, a cost-benefit analysis must be in constant revision.

Of all the modern industrial nations of the earth, the Japanese have had the most ambitious, farsighted, and directed program to develop the new ceramics. Figure 3.1 shows the areas that have been selected for concentration by Japan. The Japanese are noted for their policy of buying technology and patents from all who have

31

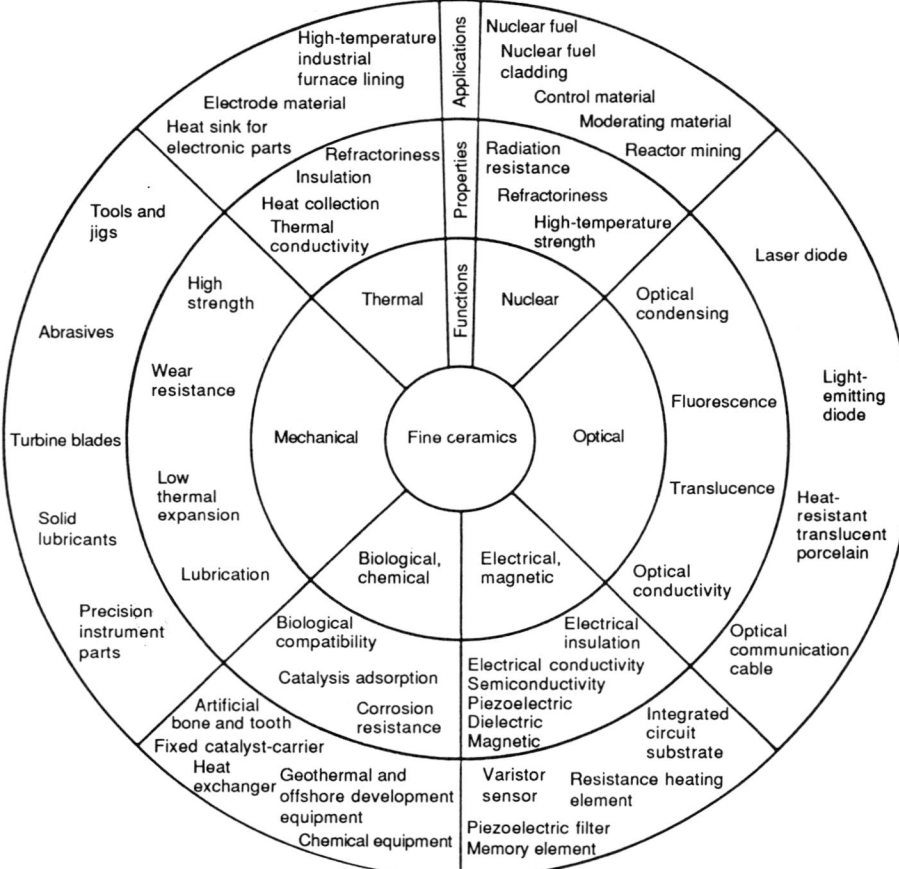

Figure 3.1 Examples of functions and applications of fine ceramics. (Courtesy of the Fine Ceramics Office, Ministry of International Trade and Industry, Tokyo.)

innovative concepts that Japan believes will further their own ambitious plan to lead the world of advanced ceramic applications. They then concentrate on perfecting these materials and processes for specific commercially valuable applications.

Figure 3.1 is worthy of study. The three concentric rings depict functions, properties, and applications. In the function ring one can

THE NEW CERAMICS

see the great industries that Japan has developed so successfully in the post-World War II era. For example, in the nuclear arena, Japan produces a large amount of its electrical power by nuclear means (12% in 1982) and is expanding in this direction. Their goal is to generate the equivalent of 51,000,000 kW of electricity by 1990 by nuclear means. Most of the nuclear equipment and facilities are domestically built.

The function labeled "electric, magnetic" represents Japan's huge electronic industry and all the application areas in electronic devices in which Japan has been so successful. The "mechanical" function is directed largely to the automotive area, in which Japan has excelled. One significant listing in the applications ring is the "tools and jigs" item. Ceramic cutting tools have already permitted an order-of-magnitude speedup in cutting metal, which contributes to low cost in the automotive production field. Japan has also developed the new massive type of machine tools with extremely high rigidity that are needed to use these advanced ceramic tool bits effectively. As the world, especially the United States, tries to regain lost ground in automotive competition, Japan plans to leapfrog by developing and applying new ceramic materials in it vanguard designs and processes. The chart in its entirety appears to be a blueprint for Japan's plan to develop new ceramics in order to further its lead in the postindustrial age.

Figure 3.2 is a cartoon that shows the differences between the traditional or conventional ceramics and the new or fine ceramics. The relatively simple processes on the left are to be supplanted by the complex processes on the right. This leads to applications with far more demanding requirements, represented by the rocket, nuclear reactor, turbine, and automobile. This transition is to be accomplished by control of the fine structure of the ceramic, which can only be revealed by the sophisticated instrumental means represented in this figure by the electron microscope. As an aside, Japan is already a world leader in the design and construction of such instruments. Japan is not ahead in everything. It lags in space, the aircraft industry, computer technology, energy conservation and new energy conversion devices, and the bioindustry. Japan expects that its advance in the ceramic area will help it move forward in these

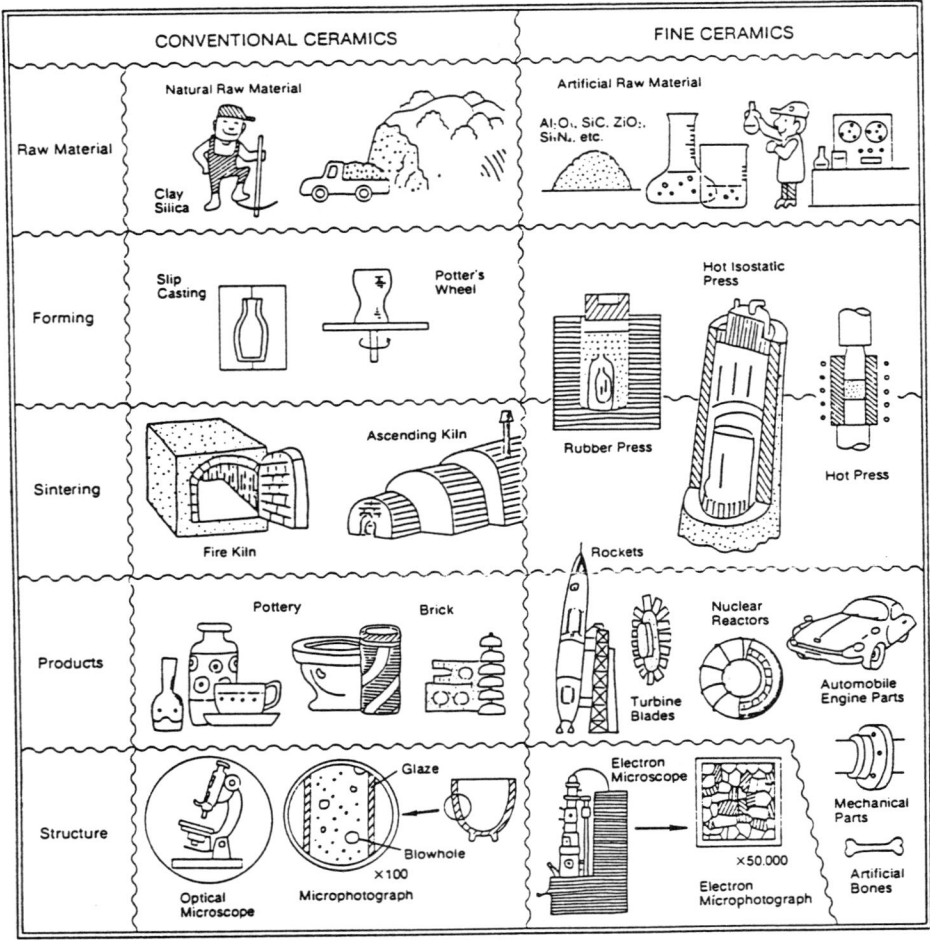

Figure 3.2 Contrast between conventional ceramics and fine ceramics. (Courtesy of the Fine Ceramics Office, Ministry of International Trade and Industry, Tokyo.)

lagging technologies. Japan's message is clear. Those who will compete successfully in the coming decades must develop and apply the new ceramics.

APPLICATIONS

Applications for the new ceramics and typical ceramics are displayed in Table 3.1. In the table only a few of the ceramic materials under development have been provided. There are a host of other ceramics in various stages of development and production. But the list as given is indicative of the range of applications and the ceramic materials available.

MECHANICAL DESIGN WITH CERAMICS

The heat engine applications listed in Table 3.1 are extremely demanding. The main difficulty in such applications is the relatively brittle nature of the ceramics compared to metal construction. The overriding advantage of the ceramics is their ability to operate at higher temperatures or at currently used temperatures without recourse to strategic metals. Higher-temperature operation translates to greater fuel economy. It is the potential for fuel economy that is the main driving force to make the huge investment needed to develop the advanced ceramics and ceramic engines that can exploit this capability.

In the United States, attention has been given to the *adiabatic diesel engine*, where the ceramic components act as thermal insulation, avoiding the need for water cooling. The elimination of water cooling of the structural parts means that less energy is lost and the engine can operate more efficiently. The second great advantage is that no coolant system is required, thus eliminating the maintenance associated with the fluid, pump, and radiator. For military vehicles, both maintenance and vulnerability of the engine to hostile action are reduced. Photographs of several of the ceramic components are shown in Figures 3.3 through 3.6.

The second engine in development is the advanced gas turbine (AGT) engine sponsored by the U.S. Department of Energy. This is meant to be a prototype for a passenger vehicle, the main incentive

Table 3.1 The New Ceramics and Their Applications

MACHINERY
- Heat-resistant structural materials
  Si_3N_4
  SiC
  Cordierite ($2MgO \cdot 2Al_2O_3 \cdot 5SiO_2$)
  Mullite ($Al_2O_3 \cdot 2SiO_2$)
  ZrO_2 (transformation toughened)
- Heat engine components
  Diesel engine
    Cylinder liner
    Piston cap insulator
    Valve seats
    Valve guides
    Valve seats
    Glow plugs (Si_3N_4)
  Turbine engine
    Inlet guide vanes
    Exhaust diffuser
    Turbine wheel with integral blades
    Turbine blades
    Regenerator
    Combustor
  High-temperature bearings
  High-temperature transport roller
  High-temperature gas and molten metal transport pipe
  Stirling engine (external combustion engine) components
  Ceramic heat exchanger (cordierite)
- Wear-resistant materials
  Mechanical seats (SiC)
  Thread guides (Al_2O_3)
  Ceramic liners (ZrO_2)
  Coated wear surfaces (TiC)

CUTTING TOOLS
  WC (conventional)
  Al_2O_3
  $Al_2O_3 \cdot TiC$
  ZrO_2-toughened Al_2O_3
  TiC

THE NEW CERAMICS

Table 3.1

TiC·TiN
Si_3N_4
SiAlON
TiN cermet
Cubic BN
Manufactured diamond
WC coated with TiC, TiN, cubic BN,
   or manufactured diamond

EMISSION—CONTROL CATALYST CARRIERS

Honeycomb cordierite

NUCLEAR

- Fuel
  UO_2
- Reactor components
  SiC
  B_4C
  Al_2O_3
- Dosimeters
  CaF_2
  K_2SO_4
  LiF

ELECTRONICS

- Integrated-circuit (IC) packaging and substrates
  Al_2O_3
  AlN
  BeO
  SiC
- Ceramic capacitor
  $BaTiO_3$
  $SrTiO_3$
- Thermister
  Spinel [$(NiMn)_3 O_4$, $(NiMnCo)_3 O_4$]
  $KTaNbO_3$
- Varister
  ZnO_2
- Piezoelectric
  PZT (lead zirconate titanate)
  PLZT (lead lanthanum zirconate titanate)

Table 3.1 (continued)

- LiNbO$_3$
- LiTaO$_3$
- Ferroelectric
  - BaTiO$_3$
  - Pb(TiZr)O$_3$
  - K(TaNb)O$_3$
  - LiTaO$_3$
- Ferrite
  - SrFe$_{12}$O$_{19}$
  - Y$_3$Fe$_5$O$_{12}$
- Sensors
  - Oxygen sensor (Y-doped ZrO$_2$)
  - Humidity sensor (Ti-doped MgCr$_2$O$_4$)
  - Hydrocarbon gas sensor (doped SnO$_2$)

MEDICAL AND BIOENGINEERING

- Bone replacement, artificial joints, tooth implants
  - Al$_2$O$_3$ (for implantation, wear surfaces)
  - Apatite hydroxide (for bone meal)
- Separators for biological components
- Zirconium phosphate [Zr(HPO$_4$)$_2$·H$_2$O]

OPTICAL MATERIALS

- Transparent windows
  - Al$_2$O$_3$
  - MgF$_2$
  - ZnSe
- Mirror substrates
  - SiC
  - Glass-ceramic Zerodur (70-80 wt % quartz, balance glass)
- Electro-optic materials
  - KH$_2$PO$_4$
  - Ba$_2$NaNb$_5$O$_{15}$
  - LiNbO$_3$

Figure 3.3 Zirconia valve seats.

Figure 3.4 Alumina cylinder liner.

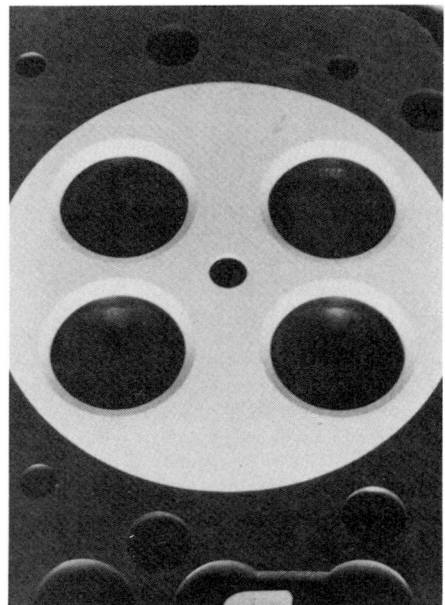

Figure 3.5 Zirconia head plate with integral seats.

Figure 3.6 NH engine zirconia-capped ductile iron piston assembly.

THE NEW CERAMICS

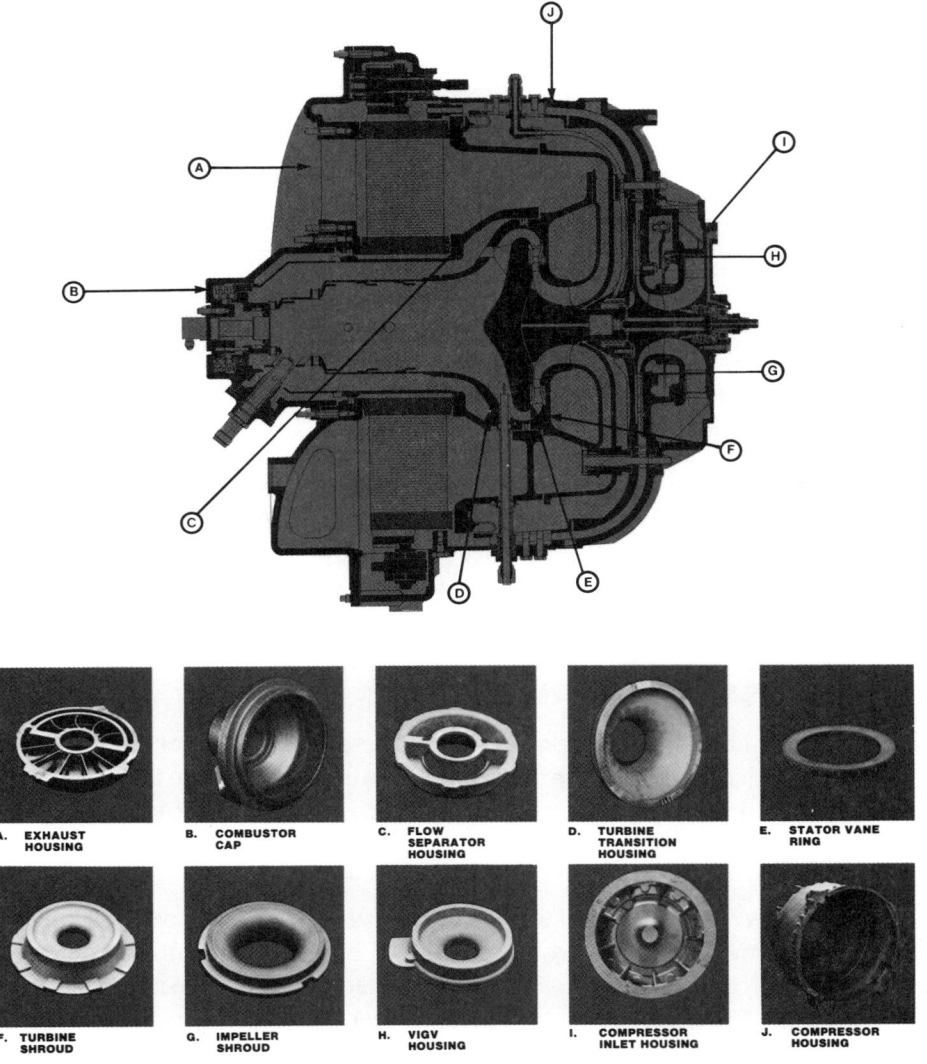

Figure 3.7 Ceramic components of the AGT 101 design that have been qualified for engine testing.

for the effort being the high-fuel-economy potential of such an engine and secondarily, a higher power-to-weight-ratio engine. The goals of the AGT 101 program conducted by Garrett/Ford are operation at 100 hp (74.6 kW) and a very low specific fuel consumption of 0.3 lb/hp-hr. The engine features a ceramic rotary heat exchanger to recover exhaust gas energy, one-piece ceramic radial ceramic turbine 15.3-cm (5.3-in.)-diameter rotating at 100,000 rpm, ceramic combustor, diffuser, and turbine shroud. Combustion gases exit the combustor at 1375 °C (2500 °F), enter the stator vanes, and then expand through the turbine to exit at 1110 °C (2030 °F) at idle prior to entering the heat exchanger to heat up the incoming air. Figure 3.7 shows various components of the engine as well as a cross section of the design.

At the conclusion of the project in 1986, the design goals had not as yet been met completely, due to temperature limitations of the aluminosilicate ceramic heat exchanger. Research on ceramic gas turbines will continue under a new proposed five-year program sponsored by the Department of Energy and NASA. In Chapter 13 we discuss this subject in more detail.

ELECTRONIC CERAMICS

Electronic ceramics have been under intensive development, and new applications are constantly being announced. A few examples are discussed here. Additional detail is provided in Chapter 12.

Ferroelectrics

Ferroelectrics are used largely for transducer applications, that is, as a means to convert from one type of excitation to another, such as from optical to electrical or from mechanical to electrical. The ceramic ferroelectric materials have been particularly useful in the developing technologies of electromechanical, electro-optical, and the acousto-optical transducers.

A ferroelectric crystal possesses an electric dipole moment even if not in an electric field. This means that in the ferroelectric state the centers of positive and negative charge do not coincide, thus in an electric field, a mechanical moment will be induced in the crystal if the field direction is not lined up precisely with the centers of charge. Above a certain temperature, called the *Curie temperature*, the ferroelectric nature of the crystal disappears.

In polycrystalline ceramics, all the electric dipole moments can be lined up by an initial application of a strong electric field at an elevated temperature followed by cooling while in the electric field. This operation is called *poling*. The ferroelectric polarization is characterized by a hysteresis loop as shown in Figure 3.8(a). The electrical setup to measure the hysteresis loop is shown in Figure 3.8(b). As the alternating field oscillates sinusoidally, the voltage across the linear capacitor, C_0, varies in accordance with the polarization of the ferroelectric, C_x. The spontaneous polarization is the value of the dipole moment at no imposed electric field. The similarity in shape of this hysteresis loop to that which occurs in ferromagnets when plotting magnetic field strength versus magnetic induction

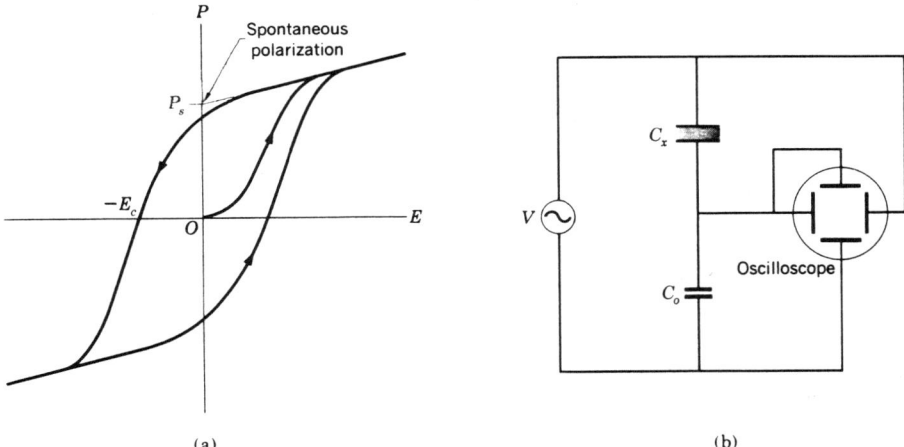

Figure 3.8 (a) Ferroelectric hysteresis loop; the coercive force is E_c. (b) Circuit for display of loop. The voltage across the ferroelectric crystal C_x is applied to the horizontal plates of the oscilloscope. The linear capacitor C_o is in series with C_x. The voltage across C_o is proportional to the polarization of C_x and is applied to the vertical plates. [After C. B. Sawyer and C. H. Tower, *Phys. Rev.* **35**, 269 (1930).]

gave rise to the term *ferroelectric*. Table 3.2 provides a more extended list of ferroelectric crystal data than that given earlier.

As mentioned earlier, these ferroelectric properties are associated with many effects that can be employed in various ways, including transducer applications and electric or electronic circuit control. Devices may use the hysteresis loop, periodic variation in capacitance of a ferroelectric capacitor, nonlinear nature of the capacitance, temperature dependency of the ferroelectric properties (pyroelectric effect), temperature dependency of the dielectric constant, electro-optical effects relating optical and electrical properties, and piezoelectric nature of ferroelectrics, described in the following section.

Piezoelectrics

Piezoelectrics are well known. These materials react to an imposed voltage by changing dimension. Thus they can be used as a way to

Table 3.2 Ferroelectric Crystal Data

		T_c (K)	P_s (esu), at T(K)
Rochelle salt group	$NaK(C_4H_4O_6) \cdot 4H_2O$	297 (upper) 255 (lower)	800
	$NaK(C_4H_2D_2O_6) \cdot 4D_2O$	308 (upper) 251 (lower)	1,100
	$LiNH_4(C_4H_4O_6) \cdot H_2O$	106	660
KDP group	KH_2PO_4	123	16,000
	KD_2PO_4	213	27,0̄00
	RbH_2PO_4	147	16,800
	RbH_2AsO_4	111	—
	KH_2AsO_4	96	15,000
	KD_2AsO_4	162	—
	CsH_2AsO_4	143	—
	CsD_2AsO_4	212	—
Perovskites	$BaTiO_3$	393	78,000
	$SrTiO_3$	32	
	WO_3	223	—
	$KNbO_3$	712	90,000
	$PbTiO_3$	763	>150,000
Ilmenite	$LiTaO_3$	—	70,000

THE NEW CERAMICS

translate voltage variations into corresponding motion. Conversely, imposing a mechanical force, such as pressure, induces a voltage generation in the piezoelectric material. Pierre and Jacques Curie discovered this effect, which they named *piezoelectricity* (Greek *piezo*, pressure) in 1880. They discovered the effect in natural materials, quartz, tourmaline (an aluminosilicate with boron), and Rochelle salt (potassium and sodium tartrate).

When a quartz crystal is precisely ground and polished to a given thickness, that crystal will oscillate at its natural frequency when a voltage oscillation at a harmonic frequency is imposed. This is the basis for the use of piezoelectric crystals in radio transmitters and receivers to tune specific broadcast frequencies. Piezoelectric tuning forks replace balance wheels in ordinary electrical wristwatches.

The great application of piezoelectrics was developed during World War II for underwater sound detection during naval operations. This procedure was termed *sonar* from "SOund NAvigation and Ranging." Barium titanate was discovered to possess the piezoelectric quality and was incorporated in early sonar equipment. From that time to the present a program has been in place to develop more effective piezoelectrics for sonar.

Ultrasonic devices, phonograph pickups, microphones, relays, actuators, and oscillating blade (bender) fans have all been developed based on the piezoelectric effect. An electrical filter is made where an electric signal is imposed on one end of a piezoelectric crystal. If the signal is resonant with the crystal, the mechanical oscillation is passed down to the other end of the crystal, where a second pair of electrodes picks up the voltage induced in the piezoelectric and transmits that signal to the rest of the circuit.

A widely used variation of this idea is the *surface acoustic wave* (SAW) delay line. In this case the fact that the mechanical wave induced in the surface of the piezoelectric crystal travels much more slowly than the electric signal would travel in a conductor permits a predetermined delay in signal propagation. Figure 3.9 illustrates such a delay device where the interdigitated electrodes can be spaced the distance apart needed to provide the desired delay. These devices are used by the millions in all sorts of electronic equipment.

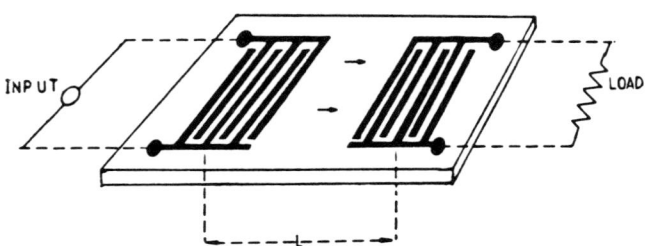

Figure 3.9 Simple SAW delay line.

Ferrites

Ferrites are the predominant class of ceramics that exhibit magnetic properties. All the ferrites are based on Fe_2O_3, the magnetic oxide of iron. They are all oxides and exhibit a magnetic induction even in the absence of an imposed magnetic field, just as permanent iron magnets do. Figure 3.10 illustrates the magnetic induction curve for a typical ferrite. Ferrites are useful because of their strong spontaneous magnetic induction, high electrical resistivity, and low loss factors. Compositions of the ferrites can readily be modified to meet specific property requirements. The magnetic properties are highly dependent on the details of the processing, such as grain size, density, and impurities.

The spinel ferrites have the same crystal structure as naturally occuring spinel, $MgAl_2O_4$. The hexagonal-structured ferrites exist in the compositional field represented by $BaO\text{-}MeO\text{-}Fe_2O_3$, where Me represents a transition metal oxide. The transition metals are those metals of the periodic table from scandium to copper. The garnet ferrites are of cubic crystal structure; $Y_3Fe_5O_{12}$, known as YIG, is the typical member of this class. The ferrites are used in many ways in electrical and electronic circuits. Table 3.3 lists various applications for MnZn and NiZn ferrites. The low-magnetic-flux-density applications are primarily signal handling. The medium- and high-flux-density applications are in the power and information-recording applications.

THE NEW CERAMICS

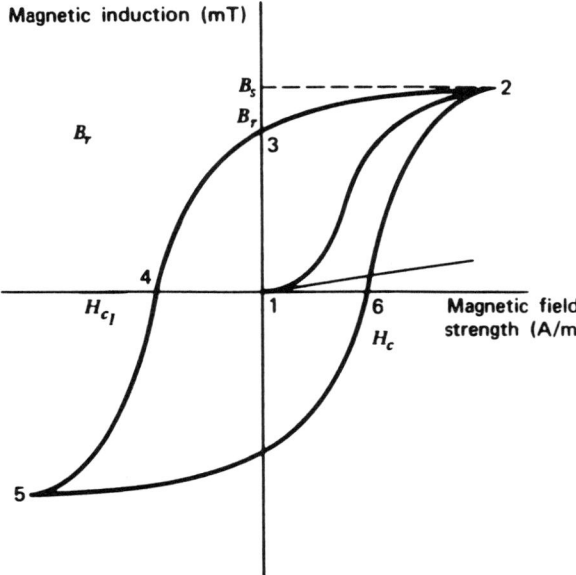

Figure 3.10 Hysteresis loop behavior for typical ferrites showing important properties. B_r, remnant polarization; B_s, saturation polarization; H_c, coercive field.

Varistors

Varistors are fabricated from ceramic semiconductors based on ZnO. The process employed to make a varistor results in crystallites of ZnO, each surrounded by a thin grain boundary in which reside elements such as Bi, Co, Pr, or Mn. The primary application of the varistor is in circuit overload protection, which is effected by the nonlinear behavior of the varistor. Such nonlinear behavior is indicated in Figure 3.11. As shown, current rises rapidly at a critical voltage, allowing circuit designs that protect from overvoltage damage. Figure 3.12 shows a typical circuit for transient protection. During normal operation the varistor acts essentially like a high-value resistor.

Table 3.3 Summary of Nonmicrowave Ferrite Compositions and Applications

Ferrite chemistry	Device	Device function	Frequencies	Desired ferrite properties
		Linear B/H, Low-Flux Density		
MnZn, NiZn	Inductor	Frequency selection network	≤1 MHz (MnZn)	High μ, high μQ, high stability of μ with temperature and time
		Filtering and resonant circuits	~1-100 MHz (NiZn)	
MnZn, NiZn	Transformer (pulse and wide band)	V and I transformation Impedance matching	Up to 500 MHz	High μ, low hysteresis losses
NiZn	Antenna rod	Electromagnetic wave receival	Up to 15 MHz	High μQ, high resistivity
MnZn	Loading coil	Impedance loading	Audio	High μ, high B_s, high stability of μ with temp., time, and dc bias
		Nonlinear B/H, Medium to High Flux Density		
MnZn, NiZn	Flyback transformer	Power converter	<100 kHz	High μ, high B_s, low hysteresis losses
MnZn	Deflection yoke	Electron-beam deflection	<100 kHz	High μ, high B_s

THE NEW CERAMICS

Composition	Part	Function	Frequency	Characteristics
MnZn, NiZn	Suppression bead	Block unwanted ac signals	Up to 250 MHz	Mod. high μ, high B_s, high hysteresis losses
MnZn, NiZn	Choke coil	Separate ac from dc signals	Up to 250 MHz	Mod. high μ, high B_s, high hysteresis losses
MnZn, NiZn	Recording head	Information recording	Up to 10 MHz	High μ, high density, high μQ, high wear resistant
MnZn	Power transformer	Power converter	<60 kHz	High B_s, low hysteresis losses
		Nonlinear B/H, Rectangular Loop		
MnMg, MnMg-Zn, MnCu, MnLi, etc.	Memory cores	Information storage	Pulse	High squareness, low switching coefficient, and controlled coercive force
MnMgZn, MnMgCd	Switch cores	Memory access transformer	Pulse	High squareness, controlled coercive force
MnZn	Magnetic amplifiers			

Note:

μ = permeability, rates of magnetic induction within the ferrite to the applied magnetic field.
Q = quality factor, ratio of field energy stored to power dissipated at a given frequency.

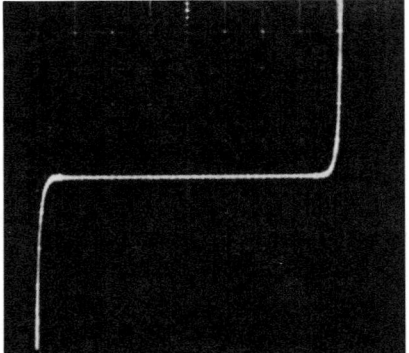

Figure 3.11 Current-voltage trace (actual photo) of a typical ZnO varistor. The applied voltage (50 V/div) is displayed on the horizontal axis and the current through the varistor on the vertical axis (200 mA/div). At a critical voltage the device shows a rapid increase in current flow.

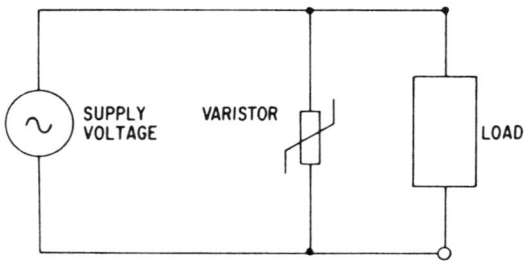

Figure 3.12 Typical application of ZnO varistor as a transient protective element.

CERAMICS IN NUCLEAR REACTORS

The fission reactor consists of fissionable fuel, moderators that control the rate of fission by absorbing neutrons, and reflectors that minimize escape of neutrons out of the reactor. Good moderators make good reflectors. Another essential element of the nuclear reactor is shielding. The shield prevents neutrons from escaping into the environment, as well as containing gamma rays produced during

THE NEW CERAMICS

Table 3.4 Advanced Ceramics for Fission Reactors

	Fuel	Fuel cladding	Moderator	Reflector	Shielding
UO_2	×				
SiC		×	×	×	×
B_4C		×	×	×	×
Al_2O_3					×

the fission process. The shielding of neutrons requires light-moderating materials, while gamma-ray shielding requires heavy materials. The various requirements of the shield can be met by the use of alternating layers, usually within a massive concrete structure. Table 3.4 lists some of the advanced ceramics currently under development for these functions. (See also Table 14.1 on reactor graphite properties.)

One of the critical problems in the development and improvement of these ceramics is to assure their resistance to damage by the nuclear environment. Upon colliding with atoms, neutrons can displace the atoms, and hence materials have to be formulated that can resist this type of degradation for long periods of time at elevated temperatures.

CERAMIC CUTTING TOOLS

The incentives to replace conventional WC (tungsten carbide) tools with ceramics are the desire to reduce dependence on strategic materials such as W, Co, and Ta and the higher cutting speeds possible with ceramic cutting tools, particularly with heavy, rigid machine tools designed specifically for use with ceramic bits. The advantages of ceramic cutting tools are high speed, ability to finish many materials that are difficult to finish with other cutters, high removal rate, superior finish, and accuracy. These characteristics of ceramic cutting tools are a consequence of the intrinsic properties of the ceramics, high hardness at room and elevated temperature, high strength in compression, good chemical stability at high temperature of operation of the tool point, and very important, high thermal conductivity, which helps to reduce tip temperature.

The main limitations of ceramics and hence the object of continuing research are lower transverse rupture strength, lower edge strength, lower fracture toughness, tendency to chip, high cost of the tools, lower uniformity of quality, abrupt failure without warning, and the need for high rigidity and therefore more expensive machining equipment. Cutting tool materials are being improved with the introduction of Si_3N_4-based ceramics, transformation-toughened ZrO_2, SiC whisker-reinforced Al_2O_3, and dense Al_2O_3-TiC materials.

In transformation-toughened ceramics ZrO_2 particles are incorporated in the matrix. The composite is heat treated in such a way that the ZrO_2 occurs in the tetragonal crystal structure. Under the influence of the strain field surrounding an advancing crack tip, the metastable tetragonal transforms to the thermodynamically stable monoclinic form with a small (about 2%) volume increase. This event effectively impedes crack propagation and enhances the fracture toughness of the material. In the SiC whisker-toughened Al_2O_3, the very fine whiskers retard crack propagation since the crack has to deviate and use additional energy to continue its advance. Cutting tools are discussed in more detail in Chapter 9.

4
Processing of Ceramics

Most of us are at least somewhat familiar with the potter's wheel, on which clay is thrown into thin-walled figures of revolution. These thrown forms are dried and exposed to an elevated temperature up to 1260 °C (2300 °F) to create a ceramic body. Conventional ceramics are natural clays, sands, and other minerals. However, in the advanced ceramic technologies, materials as obtained in nature are not sufficiently pure to permit the achievement of the properties needed. The story of the advancing front of this technology is made up of the chronicle of closer control on ingredients and processing parameters, on the development of new processes, and on improving understanding of the physics and chemistry of the processes through which the raw ingredients are transformed to ceramic bodies. There are a number of key elements to any process, as indicated in Table 4.1.

COMPOSITION

The fundamental determinant of the intrinsic properties of a ceramic is its composition. The ceramic may be unicomponent or multicomponent in nature. Alumina, Al_2O_3, is one of the most important

Table 4.1 Ceramic Process Elements[a]

Preparing the raw ingredients
Mixing the materials
Forming a 'green body,' which entails the use of a binder that is either burned out or incorporated in the finished body
Drying the green body
Green machining (optional)
Firing and densification
Finishing by grinding, cutting, polishing
Inspection (and in process quality control)

---
[a]There are many variations on these steps and many types of equipment employed.

unicomponent ceramics. The term *unicomponent* must be qualified, however, since most practical ceramics have at least small additions of binders or sintering aids that permit the consolidation of the polycrystalline starting material into a solid body.

A good illustration of the influence of composition can be obtained by considering the alumina-silica system. Silica, SiO_2, is the chemical compound constituting quartz and quartz sands. Very pure SiO_2 sands are found in various places in the world. Alumina, Al_2O_3, is obtained from bauxite rock, which contains a number of minerals, such as boehmite $AlO(OH)$, which can be converted to Al_2O_3 chemically. Corundum is a specific crystalline form of alumina.

A phase equilibrium diagram of the alumina-silica system is shown in Figure 4.1. The diagram displays the regions in the temperature-composition field that are characterized as solid, liquid, single phase, or multiple phase. For example, at about 73 wt% SiO_2, the single-phase solid substance, mullite, occurs. Mullite is an important engineering ceramic. It has the chemical representation $3Al_2O_3 \cdot 2SiO_2$. At all other compositions outside the mullite field, the ceramic will solidify from the melt into two-phase mixtures, either SiO_2 plus mullite or Al_2O_3 plus mullite. By means of compositional variation, a continuous series of useful ceramics is thereby available, all the way from 100% SiO_2 compositions to 100% Al_2O_3. Many and more complicated phase equilibria diagrams have been studied and have given rise to an understanding of a variety of complex ceramic materials.

PROCESSING OF CERAMICS

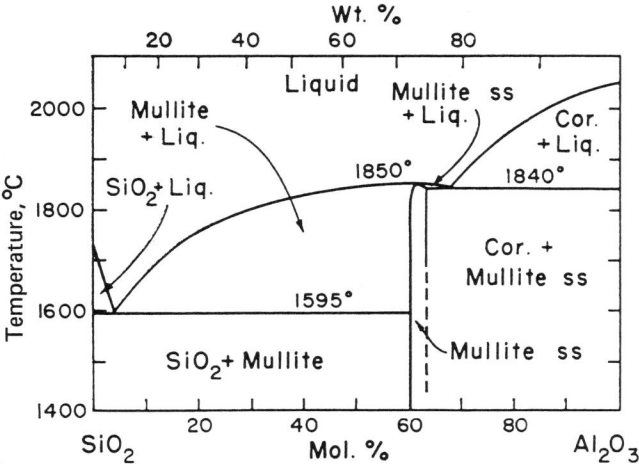

Figure 4.1 System Al_2O_3-SiO_2; revised. Cor = corundum. [From Shigeo Aramaki and Rustum Roy, *J. Am. Ceram. Soc.* **42**(12), 644 (1959); **45**(5), 239 (1962). See also N. A. Toropov and F. Ya. Galakov, *Izvest. Akad. Nauk. SSSR Otdel Khim. Nauk.* **9** (1958).]

POWDERS

Most ceramic processes start with powders of one sort or another. The powders have to have a low impurity content. Various products have different demands as to the level of purity required. Usually, the high cost of high purity imposes an economic limit on the level of purity that is affordable, even at the expense of not optimizing the properties.

The distribution of the impurity, its solubility, the details of the chemistry of the ceramic, and the property of importance or application all have a bearing on the level of impurity permissible. For example, Ca acts differently in Si_3N_4, depending on whether MgO or Y_2O_3 is used as a densification aid. Ca decreases the creep strength in the former case and is innocuous in the latter. Si_3N_4 is a major structural ceramic and is being developed for application in heat engines.

Impurities that occur as inclusions do not have much effect on creep strength, but do act as stress raisers and can cause fracture initiation at lower-than-normal levels. The degree of this effect is dependent on the relative mechanical and thermophysical properties of the inclusion and the parent material.

Particle Size

To minimize porosity in a finished piece it is necessary to have a dense green body. This is achieved by having a range of sizes in proper proportions so that tight packing of the granular material can be achieved in the green state. The powders are milled by various means to achieve near-optimum particle size distribution. It is important to avoid even a few large particles since such particles will act as defects and become areas of stress intensification under load and lead to premature failures. If the ceramic is made for its electrical, magnetic, or thermophysical properties, outsize particles may have a negative influence on performance. In addition, for such applications, which often depend on precise additions of dopants, the dopant particle size must generally be very small to improve the uniformity of distribution of the dopant through the body.

Although small particle size enhances strength, strength is not as important in the case of refractories, and relatively coarse structures are manufactured in the interest of economy. In the strongest ceramics, submicrometer particle size is desirable together with processing that develops submicrometer grain size. This goal is associated with higher cost of production. Small particles are also more reactive and enhance the densification process. For example, transparent alumina for sodium vapor lamp enclosures is made from starting particles on the order of 0.3 μm. In the case of structural Si_3N_4, bodies sintered from 2-μm average powder will sinter to only about 90% of full density, whereas with submircometer powder, where the surface area is 10 m^2/g, 95% density is achieved. Fine powders also sinter more rapidly and at lower temperatures. This provides an additional advantage in that the growth of the individual crystallites or grains is minimized. Small grain size is associated with enhanced strength.

Powder sizing is achieved by a variety of mechanisms. First, the material is broken up by some means, such as a ball mill or an attrition mill. A ball mill is simply a cylindrical ceramic jar that is

PROCESSING OF CERAMICS

rotated to mill the charge. The coarse or starting powder is placed in the jar together with solid balls or rollers, preferably of the same material as the powder so as to avoid contamination caused by ball wear. Alternatively, the material of which the balls are made is an ingredient of the composition, and the wear rate is estimated and included as part of the charge calculations. The jar also wears and needs to be considered as contributing to the composition. For a given set of conditions, the particle size distribution is a function of milling time (Figure 4.2).

Another means to comminute the powder is by means of an attrition mill (Figure 4.3). In this case the container is stationary and the grinding media are agitated by a high-speed spindel. This is a quicker method than ball milling. For either method the milling can be done dry or wet with a variety of fluids, and in any specific instance one or the other may prove able to provide a more desirable size distribution.

Other types of mills are vibratory mills and fluid energy milling where particle size reduction is accomplished by impact with high-velocity jets of fluid at sonic or near-sonic velocity. Breakup is due

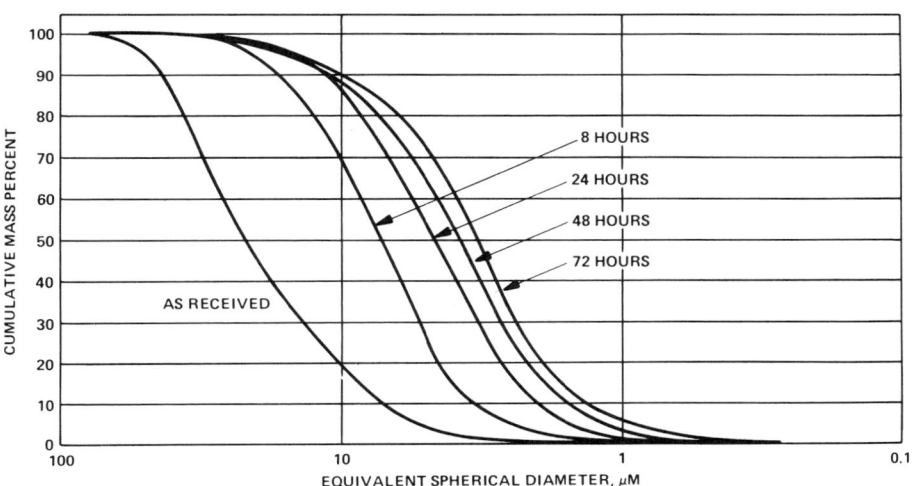

Figure 4.2 Particle size distribution of silicon powder as a function of milling time.

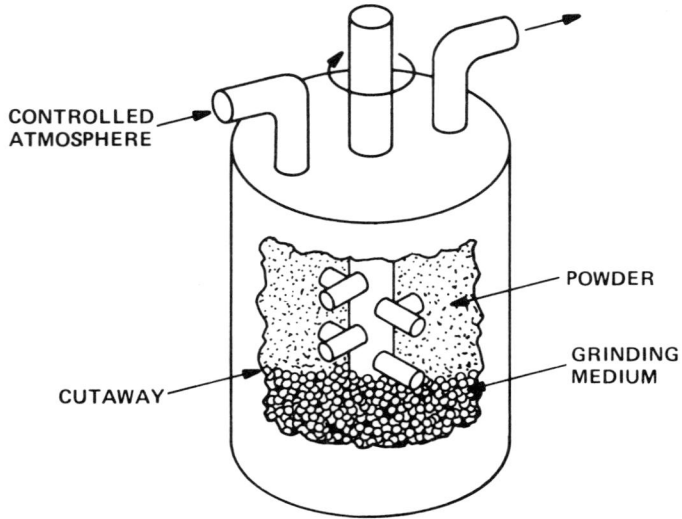

Figure 4.3 Schematic of an attrition mill. (Adapted from T. P. Herbell and T. K. Glasgow, NASA, presented at the DOE Highway Vehicle Systems Contractors Coordination Meeting, Dearborn, Mich., Oct. 17-20, 1978.)

to particle-to-particle impact. There are no moving parts and this technique tends to produce fewer contaminants. Some processes produce fine powders directly, such as precipitation, gas-phase reactions, freeze drying of soluble salts, and plasma spray.

FORMING THE GREEN BODY

In general, a green body is made from the mixture of powders appropriate to the composition desired. This can be accomplished in a variety of ways, which will be discussed a little later. Additives of various types may be added to the powders. Table 4.2 lists generic classes of additives and the purposes for which these additives are used. After the green body is formed, a drying or burnout step is employed to remove water or binders that may have been incorporated in the green body.

At some point in the drying process the green body may be machined so that after densification the part will be near the final

PROCESSING OF CERAMICS

Table 4.2 Function of Additives to Ceramics

Binder	Green strength
Lubricant	Mold release, interparticle sliding
Plasticizer	Rheological aids, improving flexibility of binder films, allowing plastic deformation of granules
Deflocculant	pH control, particle-surface charge control, dispersion, or coagulation
Wetting agent	Reduction of surface tension
Water retention agent	Retain water during pressure application
Antistatic agent	Charge control
Antifoam agent	Prevent foam
Foam stabilizer	Strengthen desired foam
Chelating or sequestering agent	Deactivate undesirable foams
Fungicide and bactericide	Stabilize against degradation with aging
Sintering aid	Aid in densification

desired (net) shape. Since machining of fired ceramics is expensive, it is desirable to obtain a fired part as near net shape as possible. The green finished shape must allow for the shrinkage that occurs during densification as the 60% dense or so green body converts to a 92% dense or so fired body. Green machining is a delicate operation due to the weakness of the green body. Tool wear is severe since the body is abrasive:

There are many ways to form the green body. The main techniques are listed in Table 4.3. Each of these techniques is sensitive to all the details of the operation. As an example, the pressing technique has many steps (Figure 4.9) and each step possesses a host of variables, which need to be carefully established and followed to produce good ware. Defects in the green body become defects in the final product.

DENSIFICATION

After producing a good green body without cracks or major pores, the next step in preparing a ceramic is the densification process. The basic process for densification is sintering. *Sintering* is the meth-

Table 4.3 Methods of Making Green Bodies

Method	Description	Advantages	Limits
Extrusion	Powder is mixed with binder to form a pliable mass and then forced by pressure through a die (See Figure 4.4)	Amenable to high production rates; low-cost tooling; used for brick, tile, tubes, rods, honeycombs (Figure 4.5)	Must be uniform in cross-section
Injection molding	Powder in a thermosetting polymeric binder is injected into a mold and hardened similarly to plastic forming; thermoset body is ejected from mold	Can make complex parts (see Figure 1.3); excellent dimensional control; high production method; see Figure 4.6 for process flow diagram	Difficult to fully remove resin without degrading green body
Pressing	Prepared powder is compressed in a die under high static force	Suitable for parts of limited size and complexity; relatively simple process	Not suitable for large or complex parts
Slip casting	A slip is prepared from water and powder; it is cast into an absorbant mold; layer of firm material adjacent to mold is removed after partial drying (Figure 4.7)	Good for large and complex parts; detail of mold is well reproduced	Slow process
Tape forming	Thin layer of raw material (slip) is laid on a flat carrier such as Teflon film; thickness of tape is controlled by a doctor blade; the slip or slurry is dried; this results in a thin tape that can be stamped into small shapes for further processing (see Figure 4.8)	High-production process for thin parts; economical	Not suitable for thick plates; process requires high degree of quality control

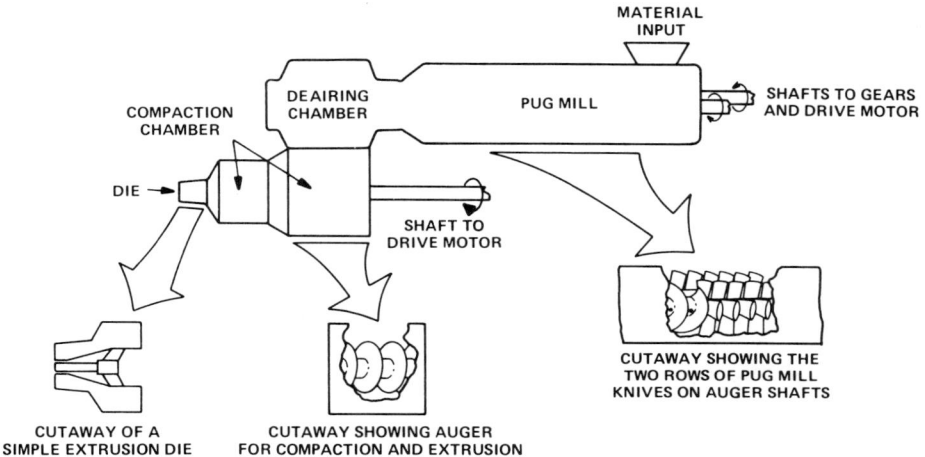

Figure 4.4 Schematic of an extruder.

od of densification by the controlled application of thermal energy to effect a bonding of the powder particles to each other with a minimum of residual porosity. As the bonding becomes more developed during the sintering cycle, the porosity decreases. Low porosity is essential for most applications for the development of high-quality mechanical or thermophysical properties.

Sintering at temperature occurs because of the transfer of matter. There are a number of transport mechanisms that may be involved, as listed in Table 4.4. Figure 4.10 depicts the diffusion process between adjacent, ideal, spherical particles. Liquid-phase sintering is a commonly employed process where a low-melting or eutectic phase forms and assists in the densification process by infiltrating the porosity and inducing various mass transport mechanisms, such as dissolution, vapor-phase transport, and precipitation. Obviously, the process is extremely complicated. It has only been in the last 50 or so years that a reasonable scientific understanding of this process has been achieved, even though sintering is an ancient technique. Some key parameters of sintering are indicated in Figure 4.11. Shrinkage, which is a measure of the removal of porosity, is shown as a function of time and temperature.

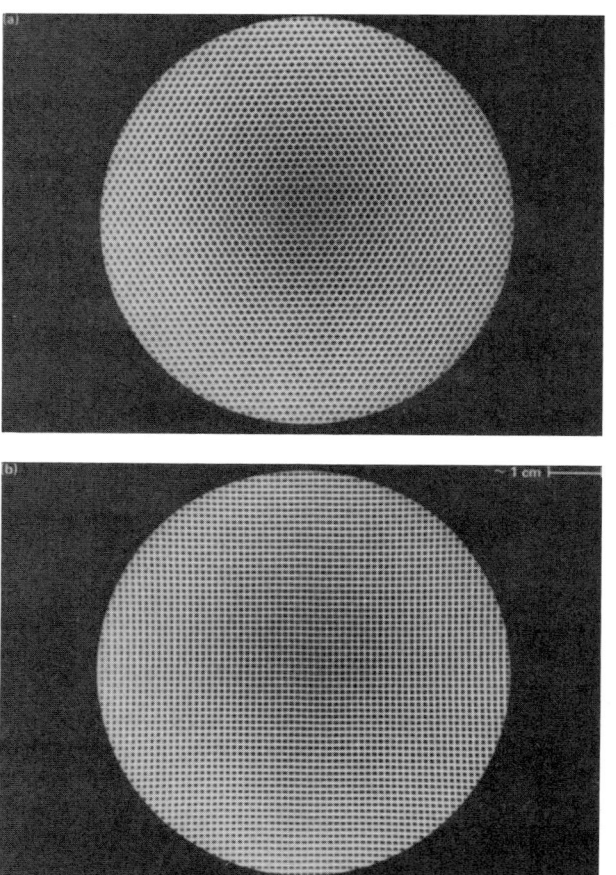

Figure 4.5 Extruded honeycomb structures for heat exchanger and emission control applications. (Courtesy of NGK Insulators, Nagoya, Japan.)

PROCESSING OF CERAMICS

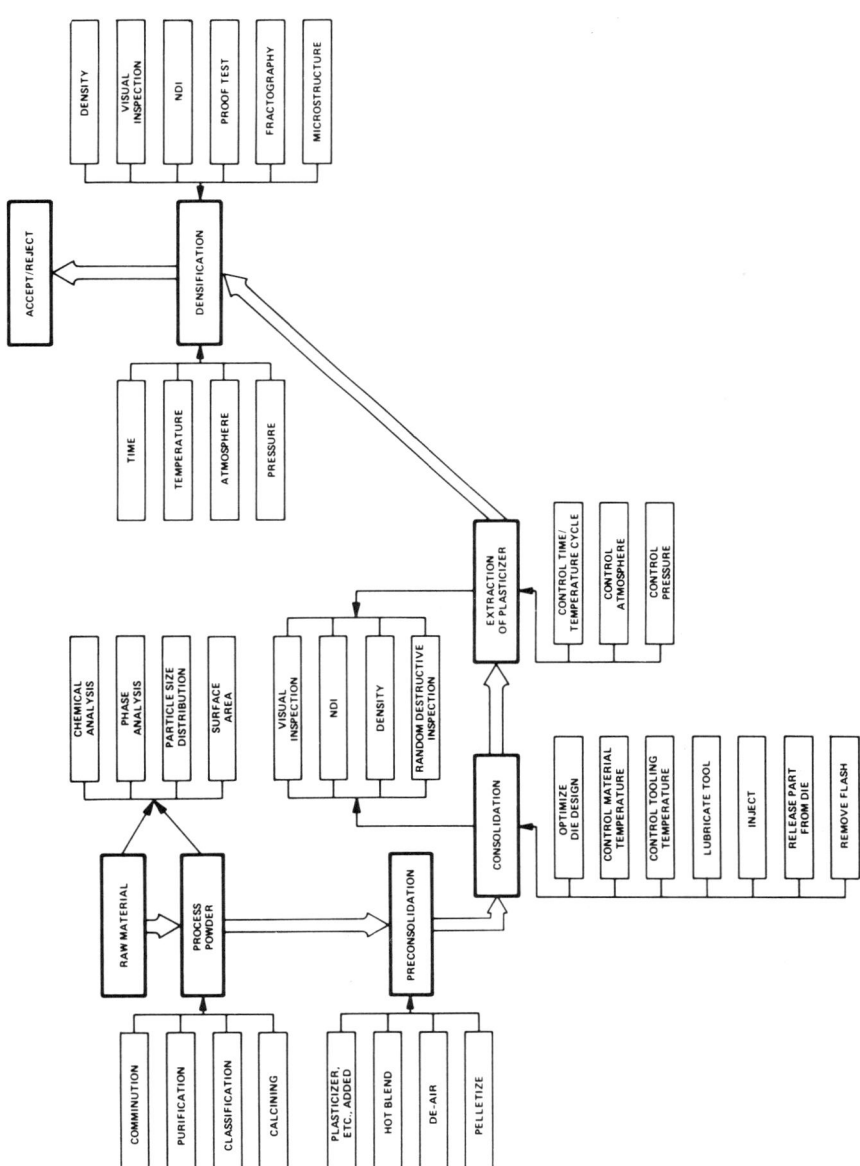

Figure 4.6 Injection molding process flow sheet.

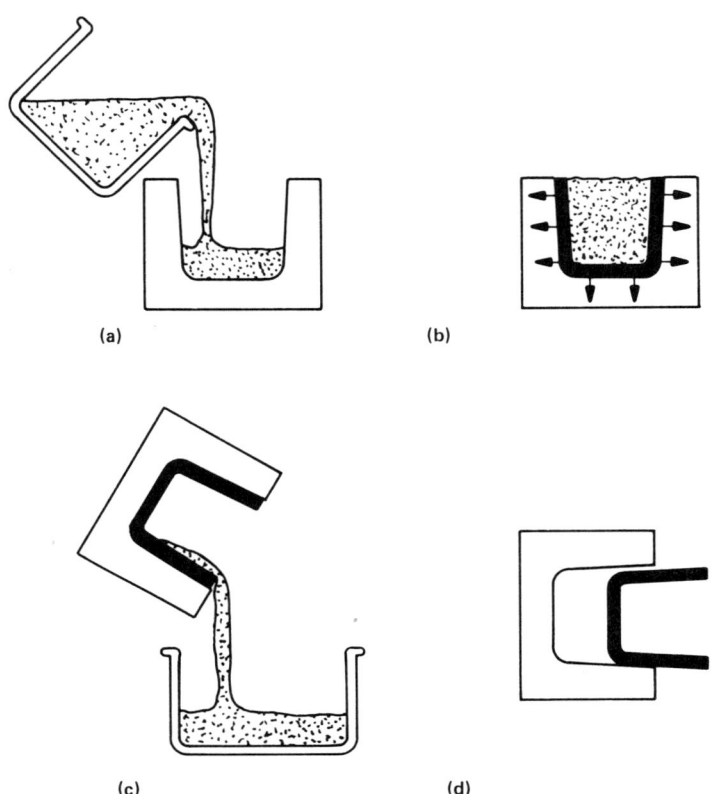

Figure 4.7 Schematic illustrating the drain-casting process.

Table 4.4 Sintering Mechanisms

Type of sintering	Material transport mechanism	Driving energy
Vapor phase	Evaporation-condensation	Differences in vapor pressure
Solid state	Diffusion	Differences in free energy or chemical potential
Liquid phase	Viscous flow, diffusion	Capillary pressure, surface tension
Reactive liquid	Viscous flow, solution-precipitation	Capillary pressure, surface tension

PROCESSING OF CERAMICS 65

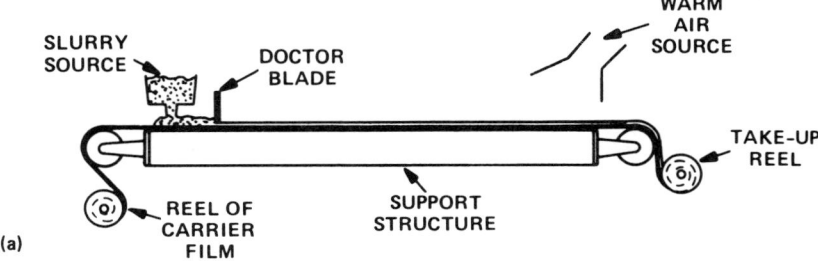

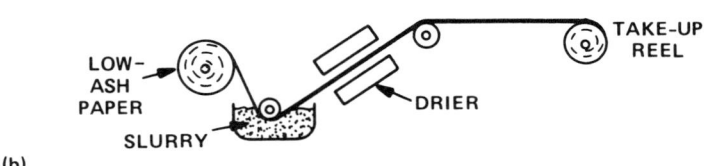

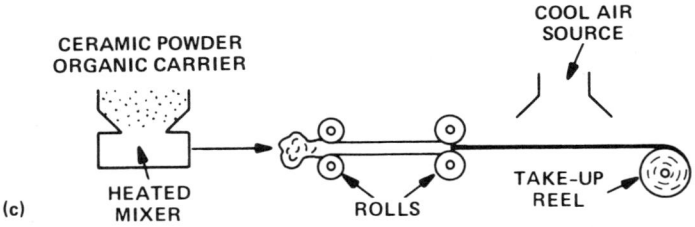

Figure 4.8 Schematics of tape-forming processes.

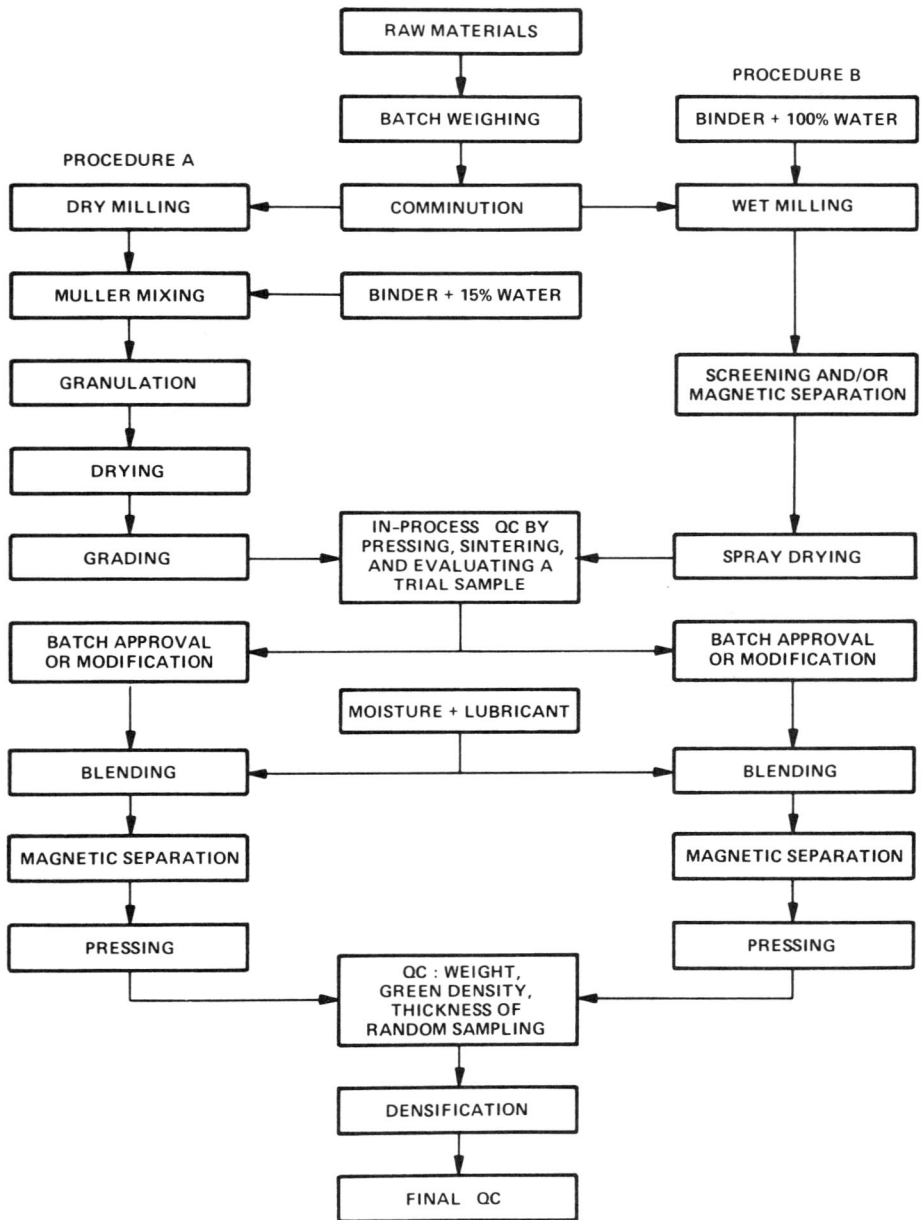

Figure 4.9 Typical flow sheets for fabrication by pressing. [Reprinted from *Ceramic Fabrication Processes* (W. D. Kingery, ed.), by permission of The MIT Press, Cambridge, Mass. ©1963, The MIT Press.]

PROCESSING OF CERAMICS

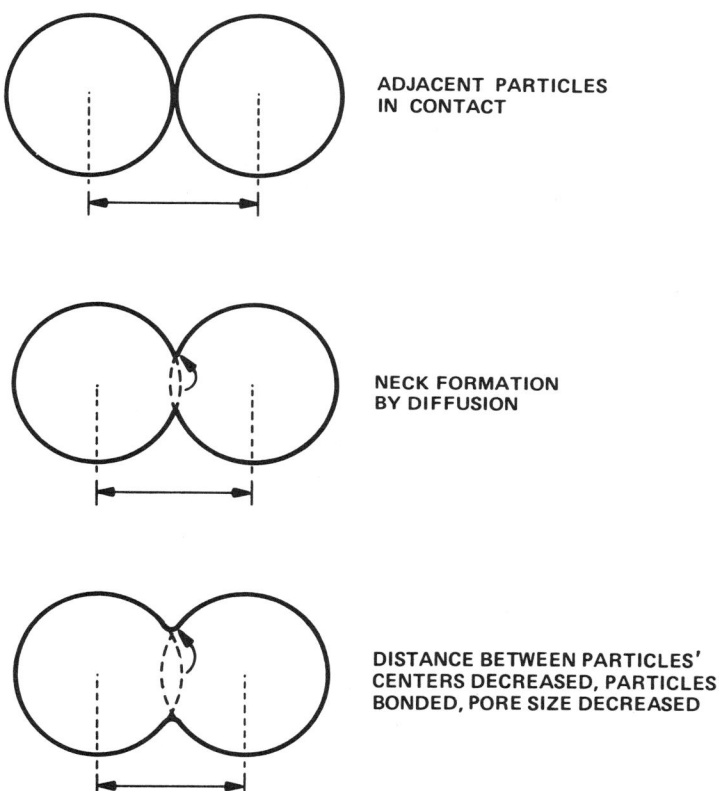

Figure 4.10 Schematic of solid-state material transport.

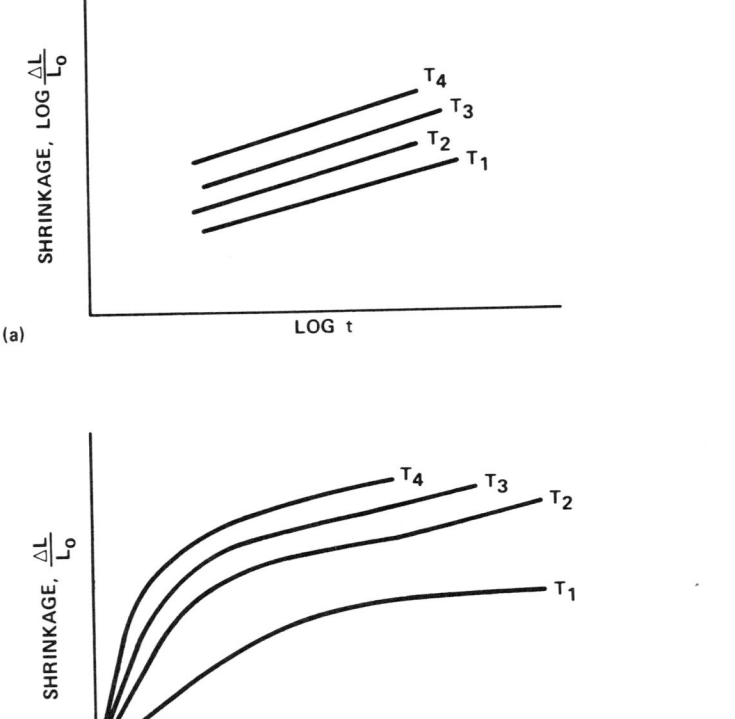

Figure 4.11 Typical sintering rate curves showing the effects of temperature and time.

Table 4.5 Alternative Densification Methods

Method	Description	Advantages	Limits
Hot pressing (HP)	Powders are placed in a die and then compressed at high pressure and simultaneously heated to sintering temperature until densified (see Figure 4.12)	Forms low-porosity bodies; high strength; minimizes size and number of flaws; reduced densification time	Limited size; expensive equipment; anisotropy; limited ability to form near net shape parts; requires protective atmosphere to avoid oxidation of dies
Hot isostatic pressing (HIP)	Powders are placed in an evacuated, sealed cannister; a pressure vessel capable of pressures up to 45,000 psi and temperatures up to 2000 °C (3632 °F) is used to heat and simultaneously, isostatically press the material by inert gas pressure until densified (see Figure 4.13); cannister may be metal or glass; presintered parts without interconnecting porosity may be HIP'ed without a container	Very dense bodies; can achieve full densification at lower temperature; permits net shape forming	Expensive equipment; high operating costs; uses high-pressure forming
Chemical vapor deposition (CVD)	Precursor gases are passed over a substrate, usually heated, and a reaction takes place on the substrate, forming the ceramic solid (see Figure 4.14 for illustration of technique)	Fine-grained structure; virtually zero porosity; very high purity	Limited to thin-walled structures or cladding; practical for simple chemistries only

(continued)

Table 4.5 (continued)

Method	Description	Advantages	Limits
Reactive sintering	A metal powder such as silicon is formed into a green body, then placed in a chamber, heated, and infiltrated with a reacting gas such as a mixture of hydrogen and nitrogen to form, in this case, Si_3N_4	Can make near-net-shaped parts; high-production method; no sintering aids, so creep strength is good	Tendency to be porous; interconnected porosity can lead to high-temperature oxidation; limited chemistries
Plasma spray	Ceramic powders are passed through a gun that incorporates electrical or combustion means to melt the ceramic powders during passage; the material is expelled at high speed; the molten stream impinges a substrate and solidifies; used most often as a coating but freestanding articles can be made on a removable mandrel	Wide range of sizes and shapes can be coated; on-site repairs can be made; applicable to many chemistries	Ceramic powders may dissociate or react with chamber environment; high-porosity parts

PROCESSING OF CERAMICS

Alternative Densification Processes

There are several processes for densification that are more involved in terms of equipment than sintering which takes place in a furnace with or without atmospheric controls. The main alternatives are hot pressing (HP), hot isostatic pressing (HIP), chemical vapor deposition (CVD), reactive sintering, and liquid particle deposition (plasma spray). Table 4.5 indicates the essence of these processes.

A wide variety of results can be achieved for even a single basic ceramic by the application of these various alternative processes. Table 4.6 illustrates the point by comparing six varieties of Si_3N_4. Users must be absolutely sure of the exact material they are speci-

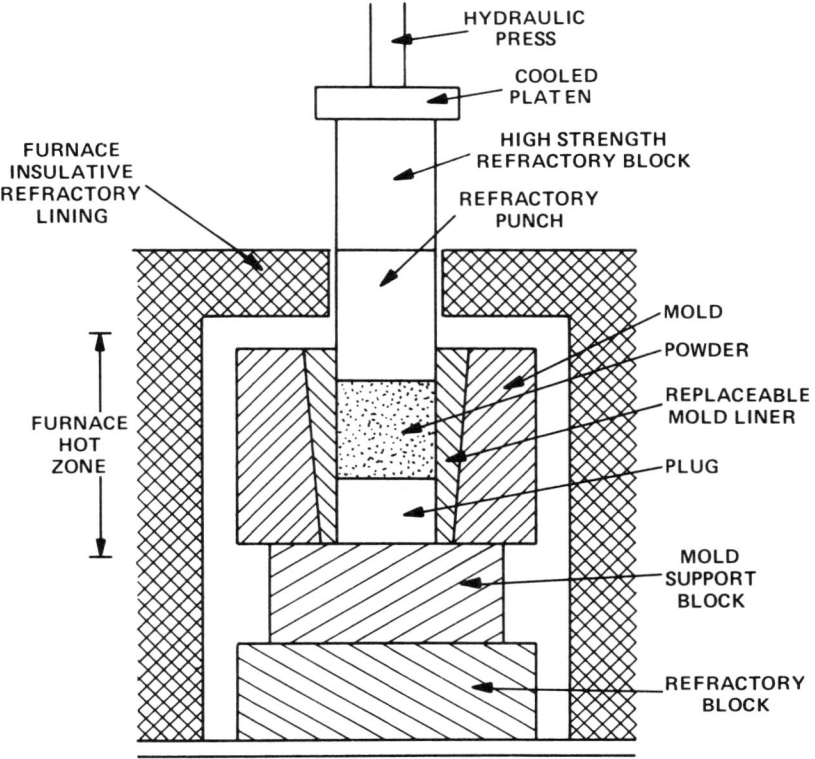

Figure 4.12 Schematic showing the essential elements of a hot press.

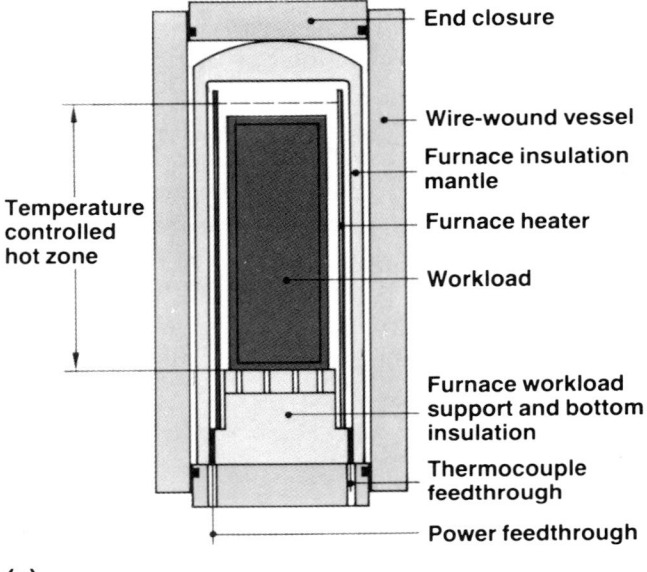

(a)

Figure 4.13 (a) Schematic of Hot Isostatic Press. (b) Principal steps in a typical HIP process. (c) Pressure/temperature cycle I. An initial preset pressure level is attained by the gas compressor. Heating is then started and the final setpoint pressure is reached as a result of temperature effects. This cycle is used for HIP'ing metal-encapsulated powders, fully densifying sintered parts, and healing defects in castings. It requires the lowest initial investment because of the smaller compressor required. (d) Pressure/temperature cycle II. Temperature is increased first, then pressurization begins. This approach requires compressors rated for the full final pressure. It is the cycle normally used for HIP'ing glass-encapsulated powders. Combinations of cycles I and II are also possible. (e) Pressure/temperature cycle III. Temperature is maintained continuously at the level desired. Pressurization is accomplished by compressors rated for the final pressure. This cycle is used in continuous production lines for products made from high-speed steel and superalloy powders. It attains the highest system utilization of the three cycles.

> These are the principal steps in a typical HIP cycle.
> 1. Workload is loaded into the press.
> 2. Press is closed.
> 3. Air is evacuated from the press by the vacuum pumping system. (This step may be repeated to assure complete purging.)
> 4. Workload is HIP'ed. Typical cycles are shown in Figs. 14c,d,e.
> 5. Pressure vessel is depressurized and inert gas is recovered.
> 6. Frame is removed and press is opened.
> 7. Workload is unloaded.
>
> The cycle can be run manually or with various degrees of automatic operation.

(b)

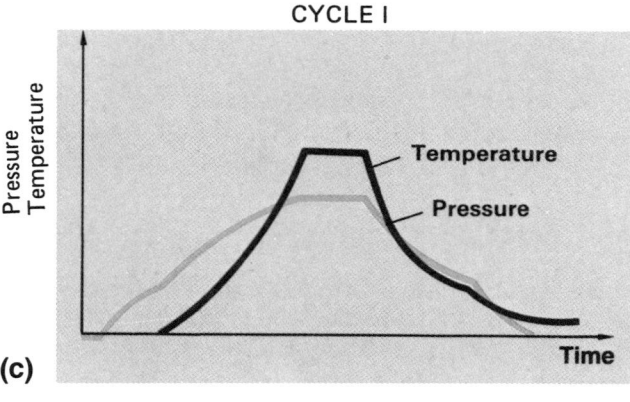

(c)

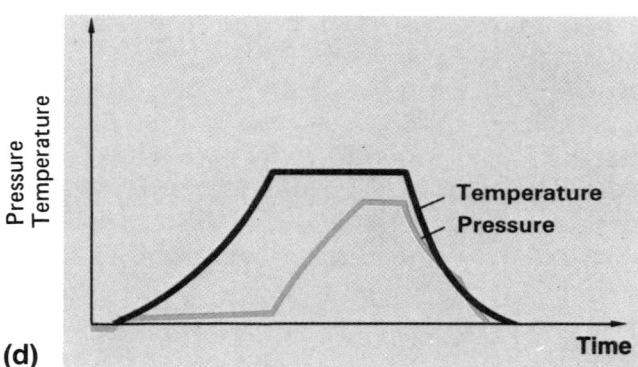

(d)

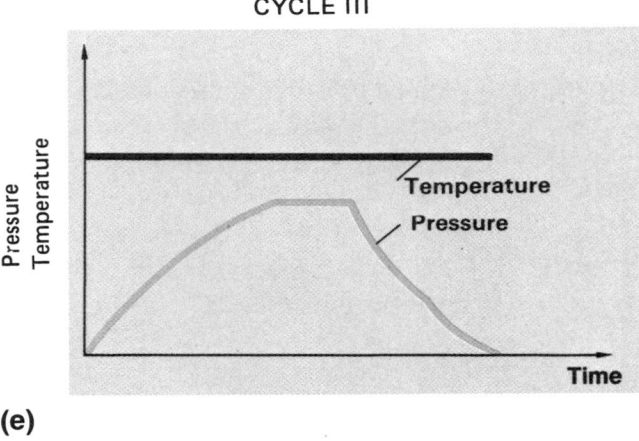

(e)

Figure 4.13 (continued)

fying and using in any specific application so that the appropriate characteristics are employed in the design and analysis phase of a project.

MACHINING OF CERAMIC BODIES

As is well understood, ceramics are brittle, hard, and prone to fracture under impact. This quality of brittleness is the most limiting characteristic of ceramics for their wide application. The vast development activity now ongoing is aimed at learning how to use ceramics despite their brittle nature. In addition, by improving their mechanical and physical properties, the degree of brittleness is reduced. The new knowledge being developed for design is also showing engineers how to apply the material reliably with reduced risk of unexpected fracture.

The problem of machining ceramics revolves around the techniques that can be used to remove material in a controllable way with a minimum of cracks and microscopic damage. The primary method for material removal is by means of abrasive machining. In fact, the observation that rough stones become smooth pebbles in the surf after rolling around in fine sand for eons was probably humankind's first realization that stone or ceramic could be shaped in a controllable and precise manner. The abrasive material, which has to be harder than the ceramic being cut, is either mounted on a

PROCESSING OF CERAMICS

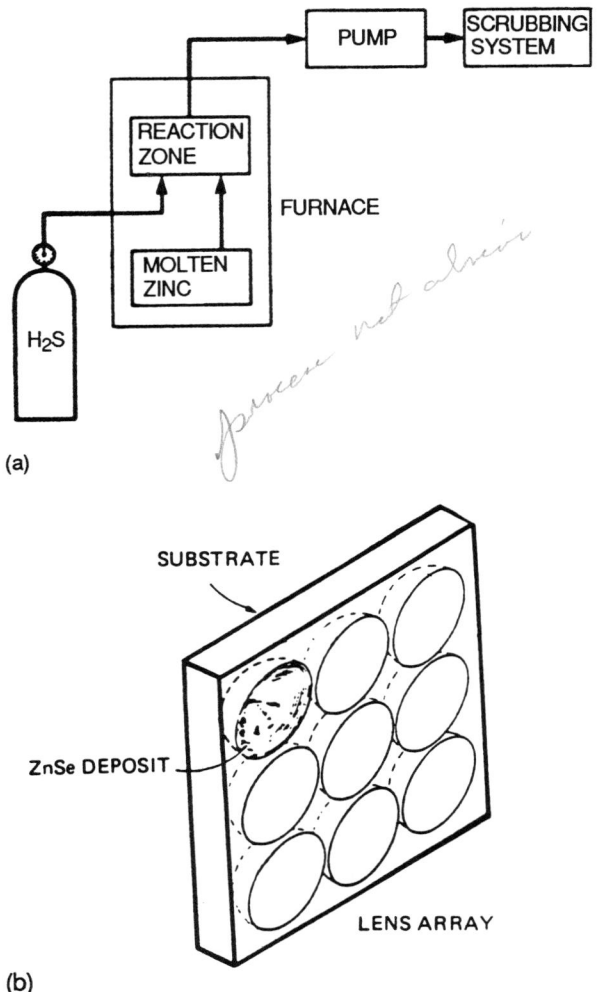

Figure 4.14 (a) Chemical vapor deposition (CVD) technique. Chemical vapor deposition process for fabrication of ZnS; the production of ZnSe is similar, but uses H_2Se rather than H_2S. (b) Mandrel for producing an array of lenses. [From R. N. Donadio, J. F. Connolly, and R. L. Taylor, New advances in chemical vapor deposited (CVD) infrared transmitting materials, *Proc. SPIE* **297**, 65-69 (1981).]

Table 4.6 Comparison of Densities and Strengths Achieved by Hot Pressing Versus Sintering

Material	Sintering aid	Density (% theoretical)	RT MOR[a] MPa	kpsi	1350°C MOR MPa	kpsi
Hot-pressed Si$_3$N$_4$[b]	5% MgO	>98	587	85	173	25
Sintered Si$_3$N$_4$[b]	5% MgO	~90	483	70	138	20
Hot-pressed Si$_3$N$_4$[c]	1% MgO	>99	952	138	414	60
Sintered Si$_3$N$_4$[d]	BeSiN$_2$ + SiO$_2$	>99	560	81	--	--
Sintered Si$_3$N$_4$[e]	6% Y$_2$O$_3$	~98	587	85	414	60
Hot-pressed Si$_3$N$_4$[e]	13% Y$_2$O$_3$	>99	897	130	669	97

[a] Room-temperature modulus of rupture.
[b] G. R. Terwilliger, *J. Am. Ceram. Soc.* **57**(1), 48-49 (1974).
[c] D. W. Richerson, *Am. Ceram. Soc. Bull.* **52**, 560-562, 569 (1973).
[d] C. D. Greskovich and J. A. Palm, U.S. DOE Conf.-791082, 1979, pp. 254-262.
[e] Data from C. L. Quackenbush, GTE Laboratories, Waltham, Mass.

PROCESSING OF CERAMICS

tool such as a grinding wheel, a free abrasive as is used in lapping, or an impacting abrasive as in sandblasting. A less commonly used technique for ceramic material removal is chemical machining such as Hf (hydrofluoric) acid for silica-based ceramics or molten sodium borate for alumina.

Another technique is *photoetching*, where a pattern is painted on a smooth ceramic surface with a photosensitive paint. After exposure to ultraviolet light through a mask and development of the pattern, the nonexposed areas can be etched into the desired pattern by means of a suitable etchant while the exposed areas form a protective film, suppressing etching in those areas. Appropriate chemistry of the substrate and various heat-treating processes are needed to permit successful application of the process, which is used widely in the electronic circuit board industry.

Electrical discharge machining can be used on ceramics that are electrically conductive. This method has been used successfully on conductive carbides, silicides, borides, and nitrides. In this process a shaped tool is held a predetermined distance from the ceramic surface and moved about by a well-controlled mechanism. A dielectric fluid circulates between the tool and the work and a high voltage is impressed, which induces sparks that essentially vaporize microscopic amounts of material, causing erosion of the surface to the desired shape.

In all these techniques microscopic damage is done to the surface and subsurface of the work. The principal function of the lapping operation is to gently remove this subsurface damage after the grosser cutting processes are completed. For this removal of defects to be successful, the entire machining process must be composed of a succession of progressively finer abrasive grit size operations. For example, rough-grind with a 200-grit diamond, finish with 320- to 600-grit diamond, rough lap with 30 to 9 μm alumina powder, and finish-lap with 3, 0.3, and finally, 0.6 μm alumina.

Amazingly, close finishes can be achieved as reported by commercial finishers and indicated in the following examples of optical finishes, which can be specified and achieved reliably for high-density, fine-grained alumina:

Parallelism to 0.000010 in. (0.25 μm)
Dimension to 0.000010 in. (0.25 μm)
Concentricity of stepped diameters to 0.000050 in. (1.25 μm)
Cylindrical inside holes between 0.060 and 2 in. to 0.000005 in. dimension (0.13 μm)

5
Structural Design Considerations

We all know that ceramics are brittle. Some ceramics are less brittle than others. In fact, some ceramics are quite tough and can take severe mechanical shocks without failure. It is the designer's task to configure the ceramic component appropriately and to select the correct material for the application. Subsequently, it is the fabricator's job to construct the component without significant flaws and to the proper material specifications. Flaw sensitivity is the hallmark of ceramics and the minimization of size and frequency of flaws and the detection of those that have not been eliminated in the fabrication cycle are essential ingredients of successful ceramic usage.

BRITTLENESS

Brittleness is not a well-defined term. There is no one property of a material that is a unique measure of its brittleness. It is, rather, a consequence of a combination of discrete, measurable properties and the environment in which the substance operates. However, brittle behavior is described as fracture with very small or no plastic flow of the material.

Figure 5.1 shows typical stress-strain curves for a brittle and a somewhat ductile material. The closer to the behavior depicted in Figure 5.1(a), the more brittle the material is. Even materials that are ductile at normal temperatures such as low-carbon steels can exhibit brittle behavior at low temperatures. The ductile-brittle transition temperature for metals is the temperature below which a normally ductile metal becomes brittle. Many old steel bridges have failed in the wintertime due to the brittle behavior of the metal in subzero weather. During World War II, many welded steel Liberty ships fractured in half during the winter months in the North Atlantic. Research showed that the impurities incorporated in the weld areas caused a severe ductile-to-brittle transition at relatively high temperature. The problem was corrected by changes in metallurgy and welding technique for subsequent constructions. Stress raisers, such as built-in notches or inadvertent internal flaws, are much more dangerous in a brittle material than in a ductile one because, in a ductile material, high stresses can cause local material yield with redistribution of stress rather than the cracking that occurs in a nonyielding brittle material.

DESIGN METHODS

In a ceramic structural element, fracture can of course be induced by impact, steady mechanical force, or by thermal stresses resulting

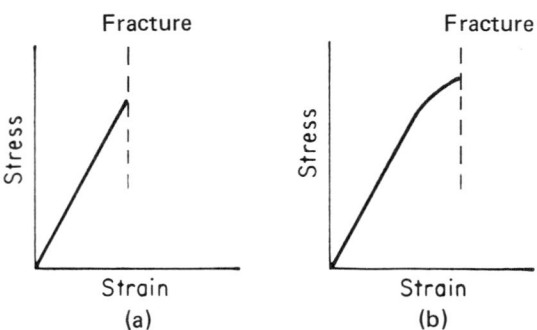

Figure 5.1 Brittle and ductile behavior.

STRUCTURAL DESIGN CONSIDERATIONS

from uneven heating of the ceramic body. In the case of steady mechanical force, point or line contact, such as on a bearing surface, can induce exceedingly high local stresses. In the case of steady-state thermal stress, well-established finite element elastic stress analysis and thermal analysis techniques exist, which can serve to provide a reasonable prediction of elastic failure. However, in the transient heating or cooling operation, these techniques become tenuous because of the difficulties associated with the exact prediction of the transient.

In addition to the thermal and stress analysis, a statistical analysis is also required to assess the probability of a given design to function successfully. The statistical analysis is a method used to take account of the relatively high variation, or scatter, in measured values of strength of ceramics over a statistical sample. This scatter is due primarily to the flaw population in the specimens, resulting in scatter of measured properties far greater than that found in ductile materials. Also, large samples exhibit lower strength than small samples because the probability of finding a large flaw is higher in a large sample. Statistical methods are needed because of the additional uncertainty concerning the positions of flaws relative to the stresses.

The measure of the scatter is a figure of merit developed by W. Weibull in 1951 called the Weibull modulus, m. The higher the value of m, the lower the amount of scatter, and consequently, the higher probability of successful operation at a given stress level. A typical Weibull distribution of test data is indicated in Figure 5.2 for a lithium aluminosilicate material. In this case 30 identical samples were subjected to a beam stress until failure. The percent of the cumulative population that failed at or before each successively higher data point is plotted against the failure stress on a log-log plot. The slope of the line is the Weibull modulus, in this case, 10. For this data set, the average bending strength was 133 MPa, with a standard deviation of 16 MPa.

The difference in size between the actual stress samples and the component being modeled must be taken into account in the statistical analysis. Based on this type of an analysis, Figure 5.3 shows the probability of failure of a typical rotor blade of a small turbine as a function of Weibull modulus and the ratio of the mean

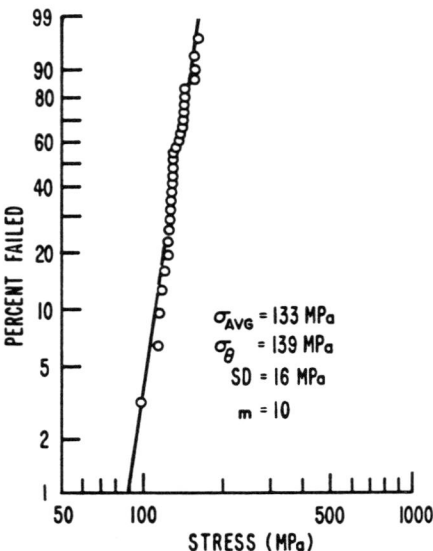

Figure 5.2 Weibull probability plot of LAS (lithium aluminosilicate).

fracture strength of the test samples to the calculated stress in the component. This ratio is akin to a factor of safety. If the mean strength is equal to the calculated stress, the probability of failure is unity in this example. For a factor of safety of 2, the probability of failure at a Weibull modulus (m) of 10 is 7×10^{-3} and for $m = 12$, the probability is reduced to 2×10^{-3}. Thus the designer has to accept some probability of failure in any practical design. The Weibull moduli used in this example are typical for high-performance, structural, monolithic ceramics. A typical plot showing the sensitivity of required mean strength to Weibull modulus is shown in Figure 5.4.

After the parts are built, proof testing of each piece is necessary in critical components to reduce the probability of failure in service to an even lower level. In the proof test, the actual component is subjected to the maximum service stress plus some arbitrary margin. The test also has to simulate the environment. The failures serve to cull out the pieces that have unacceptable and/or undetected flaws. Figure 5.5(a) shows a modified or truncated Weibull plot after proof testing has weeded out the unacceptable parts. The degree to which

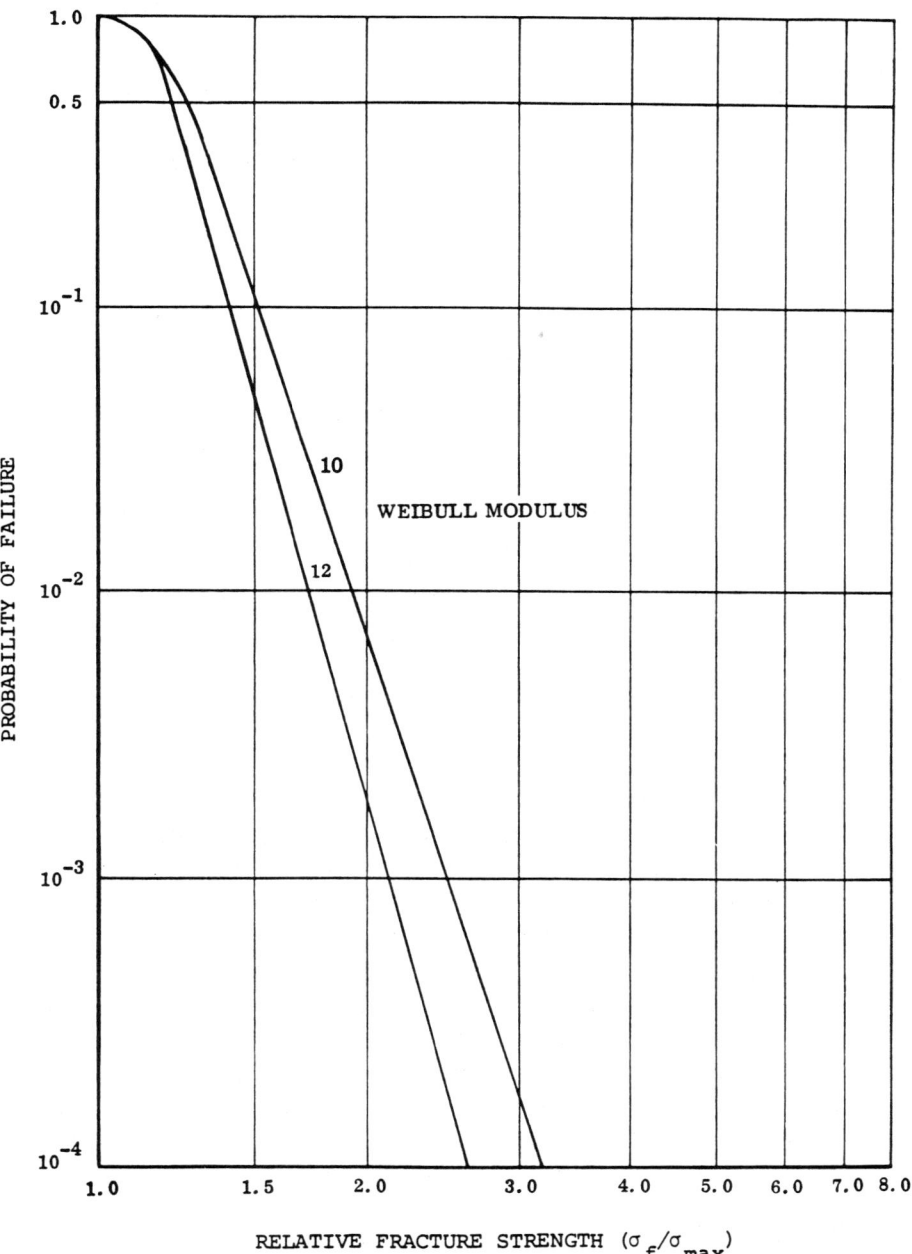

Figure 5.3 Probability of failure versus relative fracture strength.

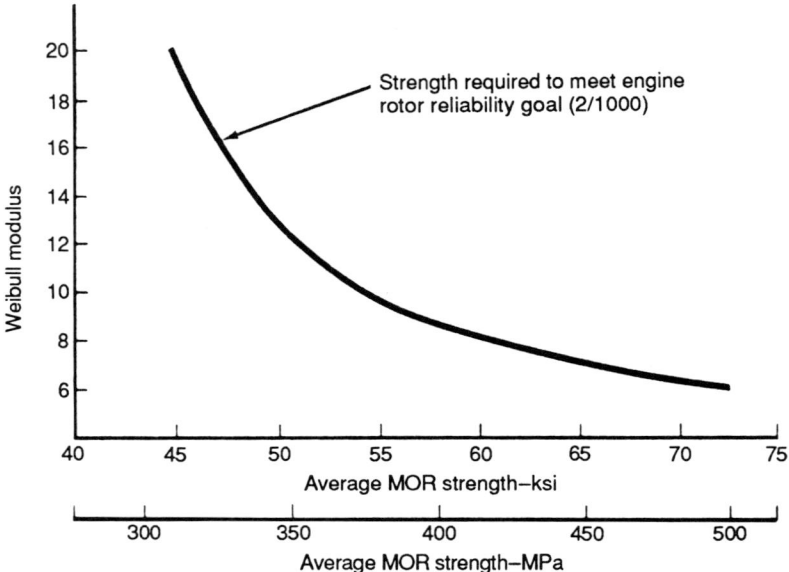

Figure 5.4 Strength and Weibull modulus requirements for CATE (ceramic automotive turbine engine) axial turbine blade.

such sophistication in design analysis and testing must be performed is a function of the value and criticality of the component. Figure 5.5(b) shows a proof-test diagram for glass windows used in the Space Shuttle. In this design, the parameter shown on the curve is the crack growth parameter, a measure of the rate of growth of the crack. This value is determined empirically. The higher the N value, the more durable is the material in the particular environment. The figure serves to define the necessary conditions of stress and N value to provide a given life for the window. Another part of the evaluation deals with the degradation of the material over time. Exhaustive characterization of the ceramic in the service environment over time has to be conducted for life predictive purposes.

Design Examples

Figure 5.6 shows injection-molded and sintered turbine rotor blades that have been formed to nearly net shape. An analysis of such a

STRUCTURAL DESIGN CONSIDERATIONS

blade includes the thermal analysis and the stress analysis. In a typical gas turbine the critical thermal condition occurs during deceleration when the thermal gradients in the blade are most severe during the initial stage of the cool-down operation. Figures 5.7 and 5.8 show a finite element model for the blade and root, and Figure 5.9 shows the maximum principal stresses for the blade. The material is SiC (silicon carbide). In this particular example, for a 45,000-rpm rotor, the maximum principal stress was computed to be about 45.6 ksi. In the normal design process, design iterations would be performed to minimize the stress levels before the statistical analysis is conducted.

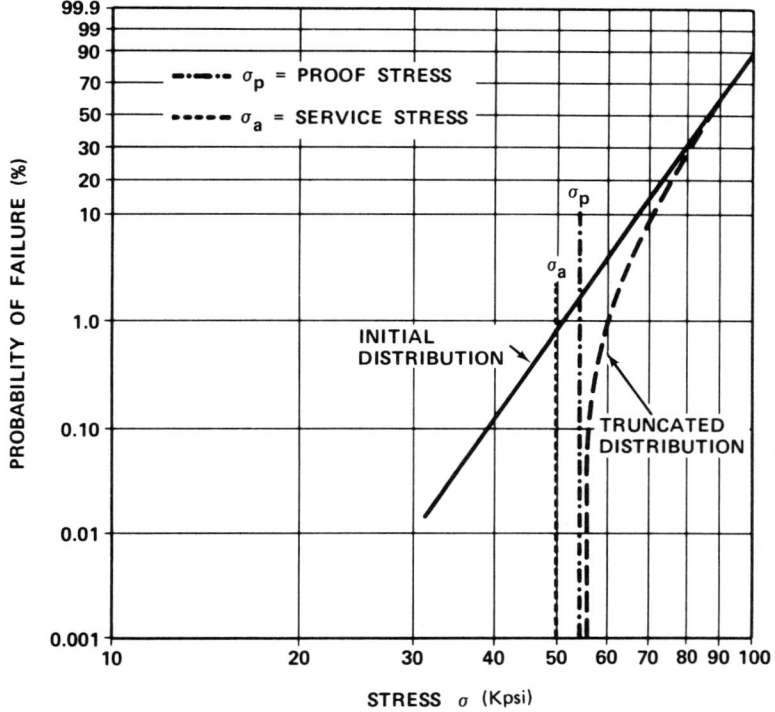

(a)

Figure 5.5 (a) Truncation of the material strength distribution to achieve improved reliability and increased operating margin. (b) Proof-test diagram for glass used in Space Shuttle windows.

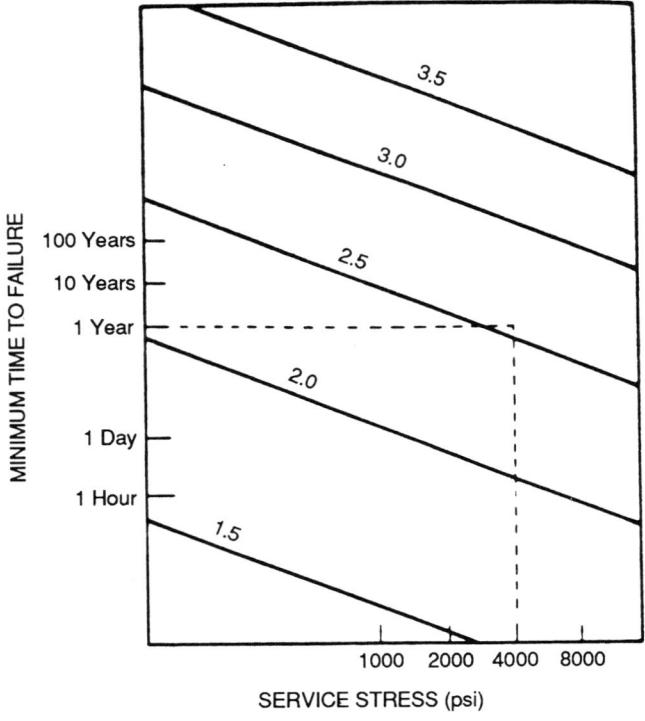

(b)

Figure 5.5 (continued)

Figure 5.6 Injection molded and sintered Si_3N_4 CATE turbine blades: (a) unoxidized; (b) oxidized 300 hours at 1000 °C; (c) oxidized 300 hours at 1200 °C.

STRUCTURAL DESIGN CONSIDERATIONS

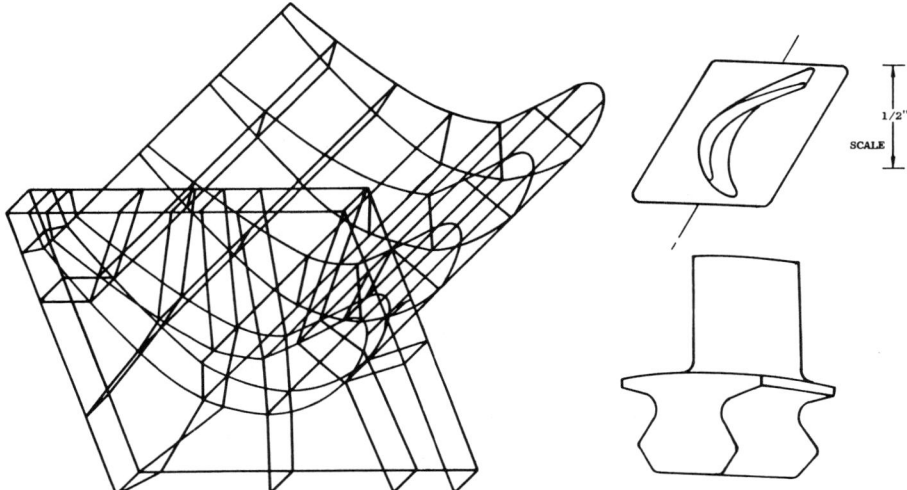

Figure 5.7 Finite element model for three-dimensional airfoil analysis.

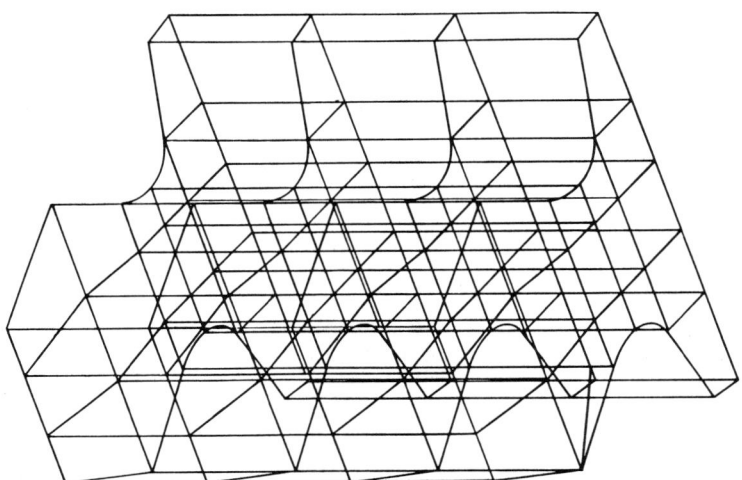

Figure 5.8 Finite element model for three-dimensional dovetail analysis.

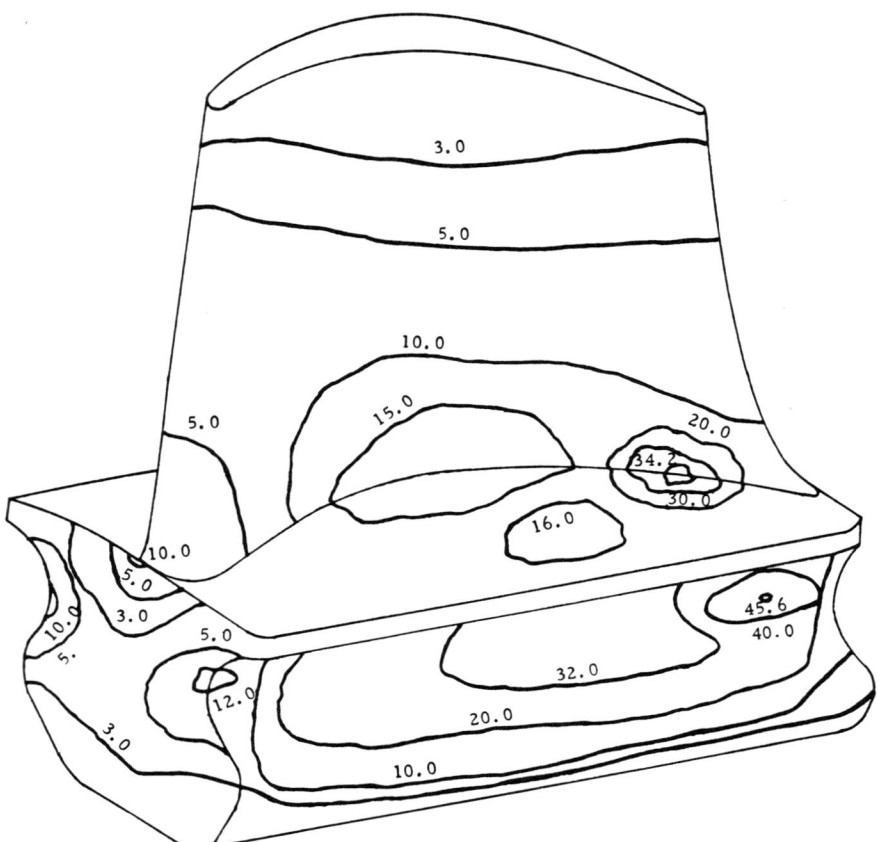

Figure 5.9 Blade maximum principal stress contour: original design.

OTHER HEAT ENGINE APPLICATIONS

A variety of heat engine ceramic component development programs are in progress driven by the goal of higher engine efficiency. The so-called adiabatic diesel engine mentioned in an earlier chapter has additional advantages, such as elimination of the liquid cooling system with an associated reduction in weight. Based on this vanguard development or perhaps parallel with it, various motor car manufacturers in the United States and overseas are working on

STRUCTURAL DESIGN CONSIDERATIONS

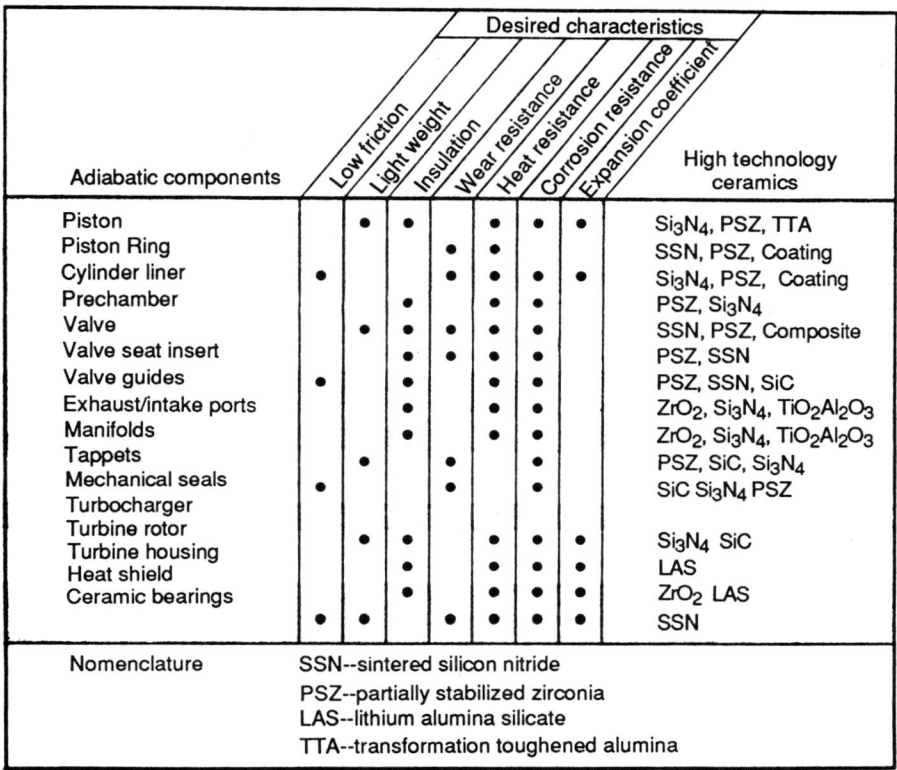

Figure 5.10 Summary of high-technology ceramics for adiabatic engines in Japan and the United States.

experimental engines that incorporate ceramic turbosupercharger wheels, cylinder liners and valve heads, ceramic piston rings, combustion chamber, exhaust ports, cylinder and piston heads, valve guides, valve seats and valves, and tappets, that is, virtually all the hot engine parts.

Figure 5.10 summarizes the advanced ceramics being evaluated for the adiabatic engine. Figure 5.11 depicts the Cummins Engine Co. AA750 adiabatic engine. Shown in Figure 5.12 is a modified T46 turbosupercharger which is being developed with various advanced ceramic components, also by Cummins. General Electric Co.,

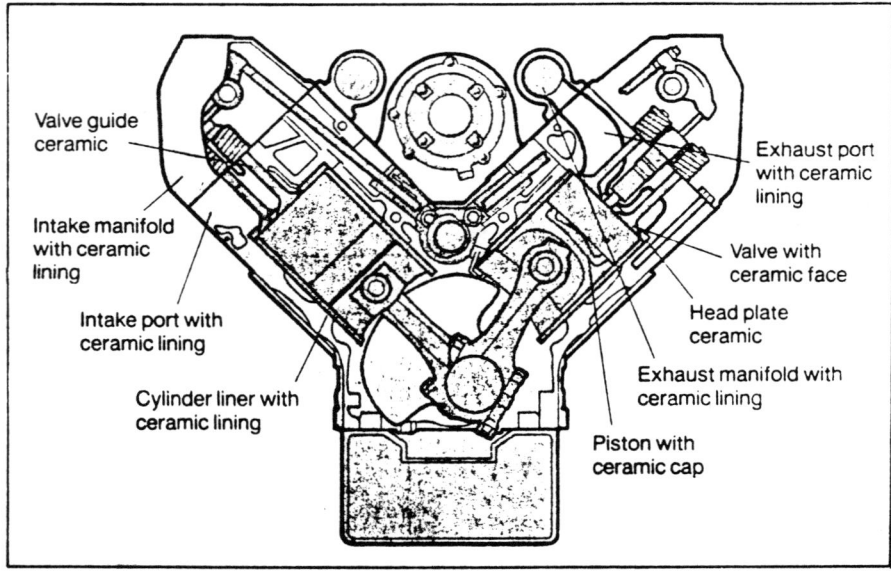

Figure 5.11 The Cummins Engine Company, Inc. and the U.S. Army Tank-Automotive Command have developed an adiabatic engine, AA750, with ceramic components. The exhaust and inlet port installation is cast "en bloc" aluminum titanate. Cylinder head and piston is insulated with partially stabilized zirconia. The cylinder liner is chrome-oxide coating over plasma-sprayed zirconia. The overall efficiency is expected to be 0.64. The first adiabatic design was tested in a U.S. Army 5-ton truck over a distance of 6000 miles and achieved a 30 to 50% increase in fuel efficiency over its production counterpart.

STRUCTURAL DESIGN CONSIDERATIONS

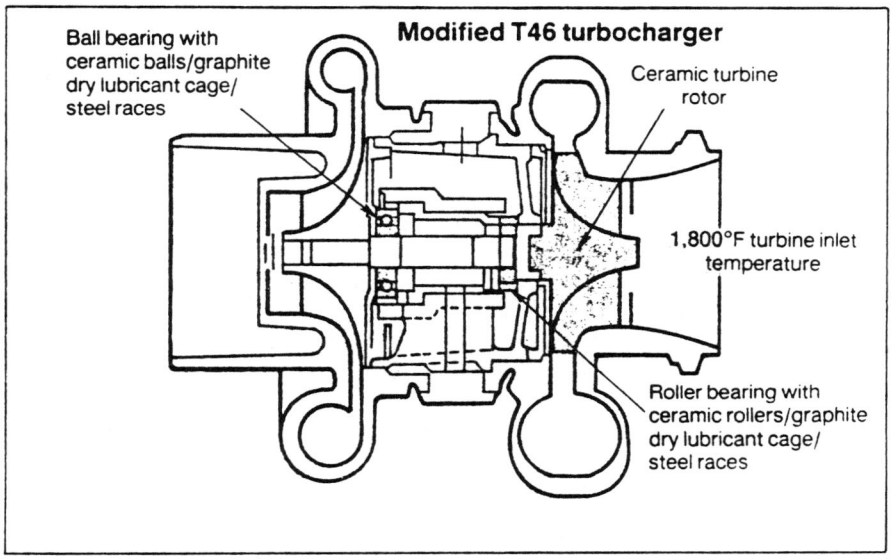

Figure 5.12 The Cummins T46 turbocharger has a pressureless sintered silicon-nitride rotor joined to a metal shaft by interference fit, based on finite element dynamic, mechanical, and thermal analyses. The ceramic turbine rotor was tested successfully in a turbocharger test cell up to 65,000 rpm at 1000 °F turbo inlet temperature for 0.25 hour. Endurance testing was also completed, with 350 hp and 1,900 rpm engine for 250 hours at 41,000 rpm and 760 °F turbine inlet temperature.

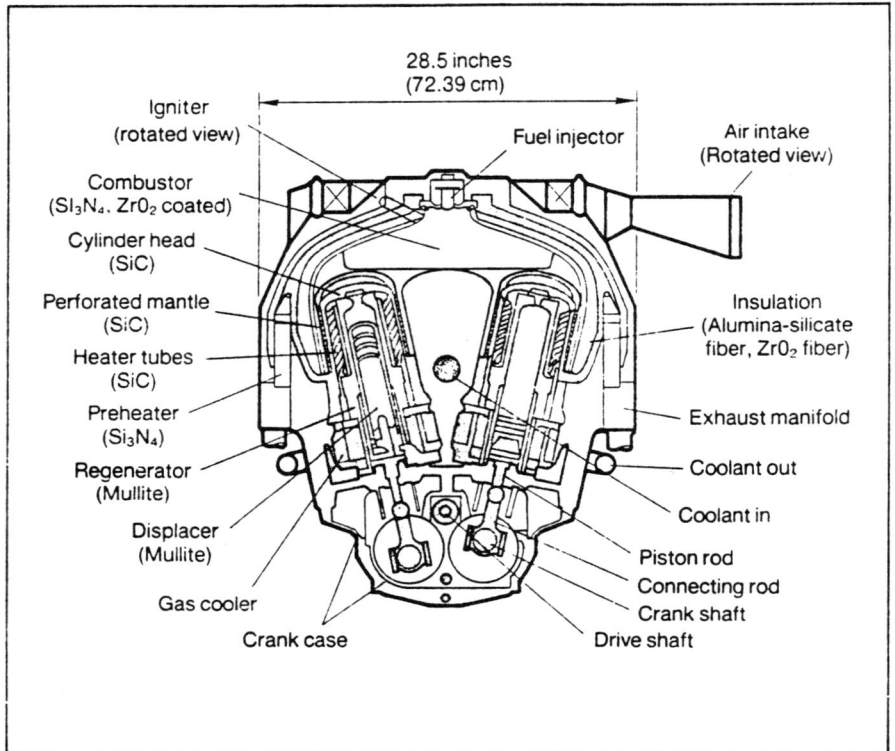

Figure 5.13 General Electric is studying the feasibility of a ceramic automotive Stirling engine (CASE) under the sponsorhip of NASA Lewis Research Center and the Department of Energy. Similar in design to the all metal Automotive Stirling Reference Engine (ASRE), which operates with hydrogen at temperatures up to 820 °C, the CASE design study selected ceramic materials for the hot section, except the outer combustor housing and engine retainer rings and bolts. The cold section below the engine cylinder and the drive system are all metal parts. The performance advantage over a metal engine is predicted to be 10 to 26% based on 1990 ceramics technology. However, several critical problems must be solved first, including joining of SiC to SiC and SiC to mullite, fabrication of thin-walled heater head tubes which must be impermeable to H_2, and development of a material for the combustor housing that can withstand temperatures up to 3800 °F. ("Ceramic Automotive Stirling Engine Study," S. Musikant, et al., January, 1985. NASA Report No. CR-174907.)

STRUCTURAL DESIGN CONSIDERATIONS

Figure 5.14 Wire drawing parts—capstans.

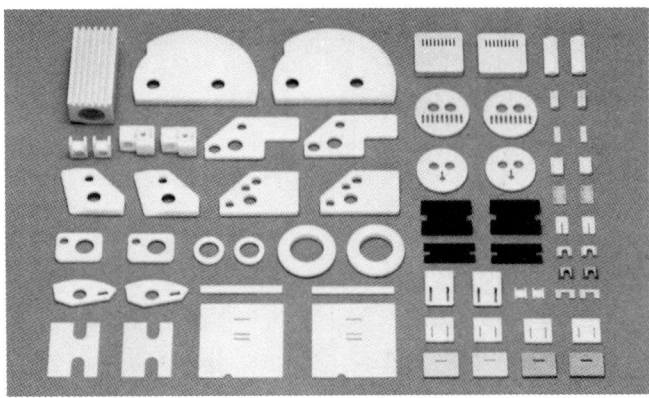

Figure 5.15 Computer peripheral parts—slider pads, tape guides, and tape cleaners.

Figure 5.16 Textile parts—friction texturizing disks.

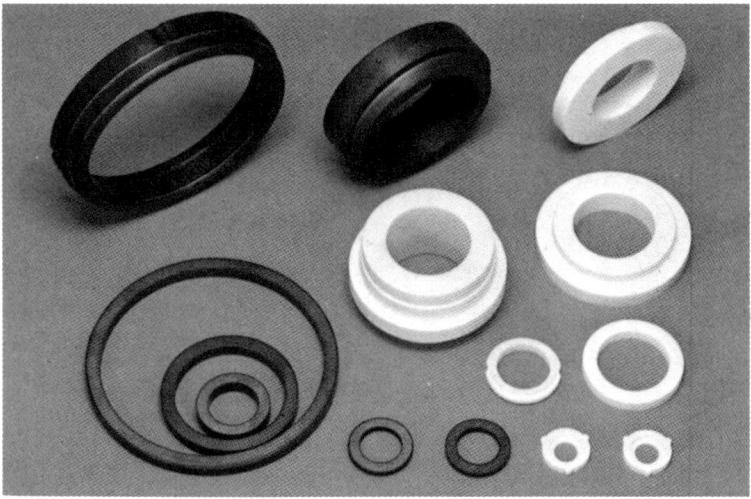

Figure 5.17a Pump parts—mechanical seal faces.

STRUCTURAL DESIGN CONSIDERATIONS 95

Figure 5.17b Balls for check valves.

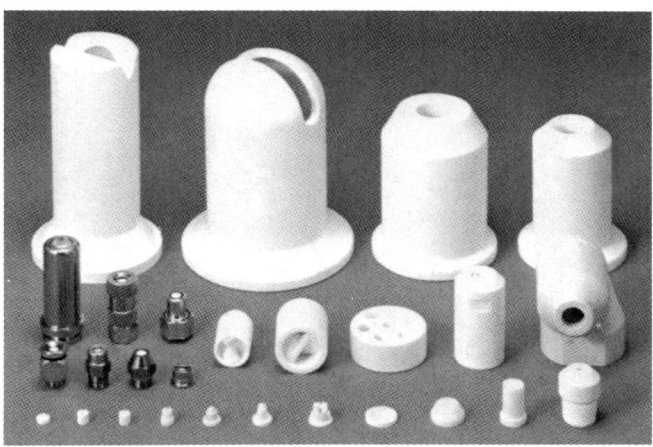

Figure 5.18 Spray nozzles for spraying abrasives, corrosive reagents, and high pressure water.

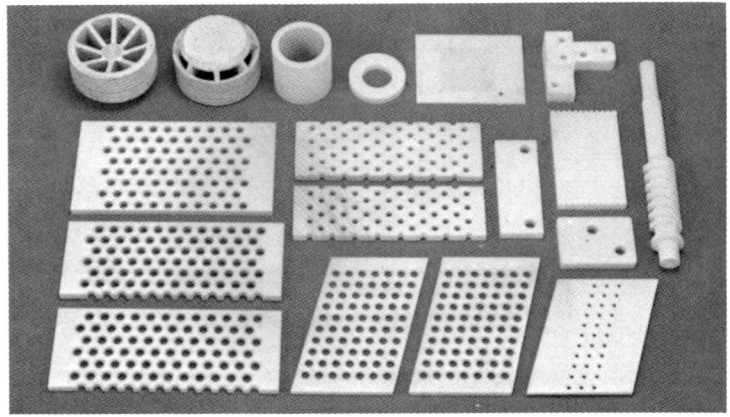

Figure 5.19 Heat-resistant nozzles for burners and glass fiber drawing.

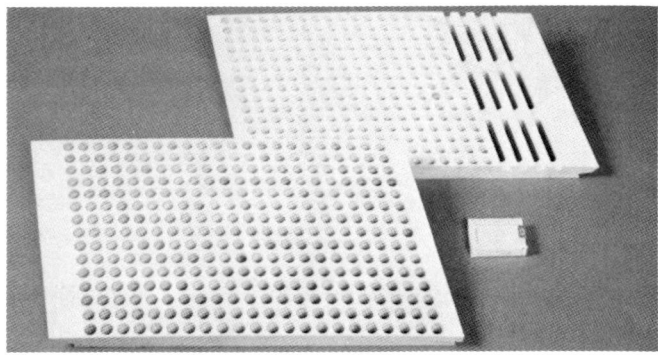

Figure 5.20a Paper industry wear parts—suction box covers.

STRUCTURAL DESIGN CONSIDERATIONS

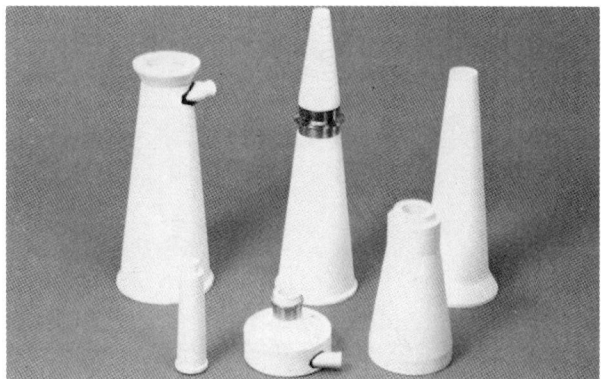

Figure 5.20b Cleaner cones.

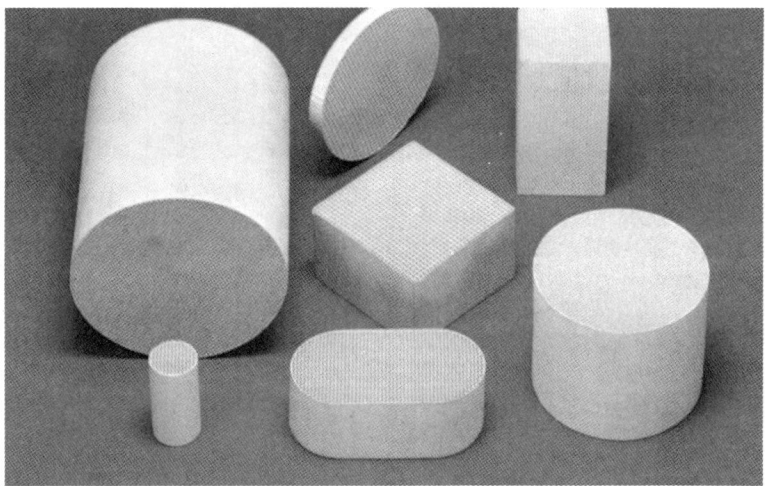

(a)

Figure 5.21 (a) Honeycomb for catalysts has miscellaneous applications.

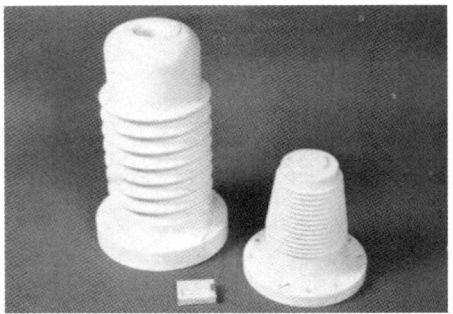

(b)

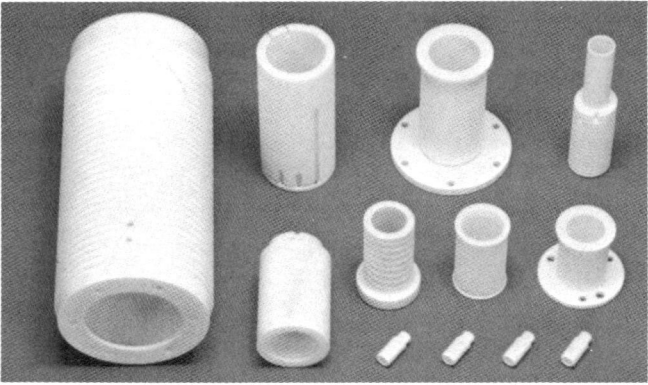

(c)

Figure 5.21 (b)Large insulators. (c) Coil forms.

under NASA sponsorship, has performed a conceptual design study for a Ceramic Automotive Stirling Engine using ceramic components. In this design, mullite was selected for the cylinder head, displacer, and regenerator, while external combustion chamber was designed using SiC. Figure 5.13 illustrates this conceptual design.

NONENGINE APPLICATIONS

A wide variety of components made from ceramics is being manufactured. Some of these items manufactured by Kyocera Corporation are shown in Figures 5.14 through 5.21.

6

Fracture Toughness

In the preceding chapter we discussed the issue of brittleness and statistical design techniques that take into account the variability of ceramic mechanical properties. In this chapter we discuss a toughness parameter used to define the maximum stress that a ceramic structural component can withstand without crack propagation. This concept assumes that the ceramic body has flaws in it prior to catastrophic failure. The fundamental concept arises from the *Griffith* (1920) *relationship*:

$$\sigma_f = \sqrt{\frac{2E\gamma}{\pi c}}$$

where

σ_f = fracture stress
E = Young's modulus
γ = energy required to create new surface (fracture surface energy)
c = flaw size prior to extension into a crack

This relationship recognizes the experimental observation that a material will withstand higher stresses if the preexisting flaw size is small and if its fracture surface energy (γ) is high.

The stress intensification factor, K, relates the geometry of the flaw and the radius at the crack tip to the average stress carried by the member. At the defect, the stresses are intensified. The primary concern in ceramics is the stress that is trying to open up the body in tension, called the mode 1 stress. Thus K is given the subscript 1 to designate this stress mode. When the flaw begins to propagate as a crack, the critical stress has been achieved and the critical stress intensification factor occurs designated as K_{1c}. The critical stress intensification factor is given by

$$K_{1c} = \sqrt{\sigma_f Yc}$$

where

Y = a dimensionless geometric factor defining the flaw

c = a crack dimension, usually length

From this equation

$$G_f = \frac{(K_{1c})^2}{Yc}$$

K_{1c} is an intrinsic property of the material, so that the larger this parameter is and the smaller the flaw size, c, the higher the critical stress required to rupture the atomic bonds and initiate a failure and the tougher the material. The units for K_{1c} are MPa·m$^{1/2}$ when stresses are expressed in MPa (megapascal) and dimensions in m (meters). Other units are MN·m$^{-3/2}$ and psi-in.$^{1/2}$.

These facts lead to the need to reduce flaw sizes in structural ceramic bodies. As a matter of fact, reducing flaw size is the major objective of ceramic process improvement developments. The goal is to reduce flaw sizes below the limit detectible by conventional nondestructive test techniques. Flaws include pores, microcracks, foreign inclusions, and inhomogeneity of chemistry or microstructure.

It follows from these relationships that as the crack grows due to fatigue or chemical reactions, the critical stress becomes lower,

FRACTURE TOUGHNESS

When the critical stress equals the actual stress at the crack tip, the material literally tears as the crack rapidly propagates through the material. The major difference in fracture behavior between the brittle ceramics and the ductile metals is the much higher value of K for the metals and the consequently much higher value of the critical stress needed to cause a catastrophically rapid crack growth in the ductile metals. Table 6.1 is an excellent summary of properties of common ceramic materials. Note the range of values for fracture toughness, from values 0.75 MPa·m$^{1/2}$ for Pyrex to 18 for cemented carbides. In comparison, cast iron has fracture toughness in the range 37 to 45. Since we think of cast iron as a brittle metal, the extremely greater brittleness of ceramics is evident by comparing these fracture toughness data. Figure 6.1 illustrates the state of stress in a body under lateral tension as indicated. The stresses are highest at the crack tip, lower at the elliptical flaw major axis terminus, and still lower between the discontinuities.

These ideas can easily be illustrated with ordinary typing paper. Prepare two sheets of paper as follows:

Sheet 1: As is.
Sheet 2: Cut 1/2 in. into the long edge midway between top and bottom.

Grasp sheet 1 at the corners of the long edge. Pull, gradually increasing the force until the sheet rips. Repeat with sheet 2 with the cut on the edge you are grasping. You will find that the defect induced in sheet 2 has greatly reduced the force needed to produce a failure. In other words, c, the flaw size, has been made much larger, and therefore the critical stress needed for failure has been reduced.

MEASUREMENT OF FRACTURE TOUGHNESS

The stress intensification factor, K_{1c}, or fracture toughness is measured by highly controlled fracture experiments. The most common and easiest test is an indentation test where a sample of material is subjected to the force of a pyramidal diamond indentor (Vickers test) until the material cracks. For well-behaved specimens, four cracks occur at the edges of the pyramidal indent. The length of these

Table 6.1A Properties of Heat Engine and Common Ceramic Materials

Material	Crystal structure	Theoretical density (Mg/m^3)	Knoop or Vickers hardness (GPa)	Transverse rupture strength (MPa)	Fracture toughness (K_{1c}) (MPa·m$^{1/2}$)
Glass ceramics	Variable	2.4-5.9	6-7	70-350	2.4
Pyrex glass	Amorphous	2.52	5	69	0.75
TiO$_2$	Rutile tetragonal	4.25	7-11	69-103	2.5
	Anatase tetragonal	3.84			
	Brookite orthorhombic	4.17			
Al$_2$O$_3$	Hexagonal	3.97	18-23	276-1034	2.7-4.2
Cr$_2$O$_3$	Hexagonal	5.21	29	>262	3.9
Mullite	Orthorhombic	2.8		185	2.2
Partially stabilized ZrO$_2$	Cubic, monoclinic, tetragonal	5.70-5.75	10-11	600-700	8-9 at 293 K, 6-6.5 at 723 K, 5 at 1073 K
Fully stabilized ZrO$_2$	Cubic	5.56-6.1	10-15	245	2.8
Plasma-sprayed ZrO$_2$	Cubic, monoclinic, tetragonal	5.6-5.7		6-80	1.3-3.2
CeO$_2$	Cubic	7.28			

FRACTURE TOUGHNESS

Material	Structure				
TiB$_2$	Hexagonal	4.5-4.54	15-45	700-1000	6-8
TiC	Cubic	4.92	28-35	241-276	
TaC	Cubic	14.4-14.5	16-24	97-290	
Cr$_3$C$_2$	Orthorhombic	6.70	10-18	49	
Cemented carbides	Variable	5.8-15.2	8-20	758-3275	5-18
SiC	α hexagonal	3.21	20-30	Sintered	Sintered
	β cubic	3.21		96-520 at 300 K	4.8 at 300 K
				250 at 1273 K	2.6-5.0 at 1273 K
				Not pressed	Not pressed
				230-825 at 300 K	4.8-6.1 at 300 K
				398-743 at 1273 K	4.1-5.0 at 1273 K
SiC (CVD)	β cubic	3.21	28-44	1034-1380 at 300 K	5-7
				2060-2400 at 1473 K	
Si$_3$N$_4$	α hexagonal	3.18	8-19	Sintered	Sintered
	β hexagonal	3.19		414-650	5.3
				Not pressed	Hot pressed
				700-1000	4.1-6.0
				Reaction bonded	Reaction bonded
				250-345	3.6
TiN	Cubic	5.43-5.44	16-20		
Graphites (with grain)	Hexagonal	2.21	35-85[f]	0.48-207	0.5-1.8
Cast irons	Cubic	5.5-7.8	1.7	90-1186	37-45

Table 6.1B Properties of Heat Engine and Common Ceramic Materials

Material	Young's modulus (GPa)	Poisson's ratio	Thermal expansion ($n\ 10^{-6}\ K^{-1}$)	Thermal conductivity [W/(m·K)]	Specific heat [J/(kg·K)]	Emittance[b]	Thermal shock resistance parameter[c]
Glass ceramics	83-138	0.24	5-17	2.0-5.4 at 400 K 2.7-3.0 at 1200 K	795-1298	0.9 at 300 K (T)	1.2[d]
Pyrex glass	70	0.2	4.6	1.3 at 400 K 1.7 at 800 K	335 at 100 K 1170 at 700 K	0.85 at 100 K (N) 0.85 at 900 K (N) 0.75 at 1100 K (N)	0.2
TiO_2	283	0.28	9.4	8.8 at 400 K 3.3 at 1400 K	799 at 400 K 920 at 1700 K	0.83 at 450 K (T) 0.89 at 1300 K (T)	0.2
Al_2O_3	380	0.26	7.2-8.6	27.2 at 400 K 5.8 at 1400 K	1068	0.75 at 100 K (N) 0.53 at 1000 K (N) 0.41 at 1600 K (N)	6.5
Cr_2O_3	>103		7.5	10-33 at 350 K	670 at 300 K 837 at 1000 K 879 at 1600 K	0.69 (N) 0.91 (N)	2.7
Mullite	145	0.25	5.7	5.2 at 400 K 3.3 at 1400 K	1046	0.5 at 1200 K (N) 0.65 at 1550 K (N)	0.9
Partially stabilized ZrO_2	205	0.23	8.9-10.6	1.8-2.2	400		0.5
Fully stabilized ZrO_2	97-207	0.23-0.32	13.5	1.7 at 400 K 1.9 at 1600 K	502 at 400 K 669 at 2400 K	0.82 at 0 K (N) 0.4 at 1200 K (N) 0.5 at 2000 K (N)	0.8
Plasma-sprayed ZrO_2	48 21 at 1373 K	0.25	7.6-10.5	0.69-2.4		0.61-0.68 at 700 K (T) 0.25-0.4 at 2800 K (T)	0.2
CeO_2	172	0.27-0.31	13	9.6 at 400 K 1.2 at 1400 K	370 at 300 K 520 at 1200 K	0.65 at 1300 K (T) 0.45 at 1550 K (T) 0.40 at 1800 K (T)	
TiB_2	514-574	0.09-0.13	8.1	65-120 at 300 K 33-80 at 1100 K 54-122 at 2300 K	632 at 300 K 1155 at 1400 K	0.8 at 1000 K (N) 0.85 at 1400 K (N) 0.4 at 2800 K (N)	21

FRACTURE TOUGHNESS

Material							
TiC	430	0.19	7.4-8.6	33 at 400 K / 43 at 1400 K	544 at 293 K / 1046 at 1366 K	0.5 at 800 K (N) / 0.85 at 1500 K (N) / 0.38 at 2800 K (N)	2.2
TaC	285	0.24	6.7	32 at 400 K / 40 at 1400 K	167 at 273 K / 293 at 1366 K	0.2 at 1600 K (N) / 0.33 at 3000 K (N)	3.7
Cr_3C_2	373		9.8	19	502 at 273 K / 837 at 811 K		0.2
Cemented carbides	396-654	0.2-0.29	4.0-8.3	16.3-119	197-544		13[e]
SiC	207-483	0.19	4.3-5.6	63-155 at 400 K / 21-33 at 1400 K	628-1046	0.85 at 400 K (N) / 0.80 at 1800 K (N)	31
SiC (CVD)	415-441		5.5	121 at 400 K / 34.6 at 1600 K	837 at 400 K / 1464 at 2000 K		
Si_3N_4	304	0.24	3.0	9-30 at 400 K	400-1600	0.9 at 600 K (N) / 0.8 at 1300 K (N)	16
TiN	251		8.0	24 at 400 K / 67.8 at 1773 K / 56.9 at 2573 K	628 at 273 K / 1046 at 1366 K	0.4 at 800 K (N) / 0.8 at 1400 K (N) / 0.5 at 2100 K (N) / 0.33 at 3000 K (N)	
Graphites (with grain)	1.4-34.5	0.07-0.22	0.1-19.4	1.67-518.8	711-1423	0.8 at 1366 K (T)	135
Cast irons	83-211	0.17	8.1-19.3	46-52	460		32[g]

Source: W. J. Lackey, D. P. Stinton, G. A. Cerny, L. L. Fehrenbacher, and A. C. Schaffhauser, Ceramic coatings for heat engine materials: status and future needs, *Proceedings of International Symposium on Ceramic Components for Heat Engines*, Oct. 17-21, 1983, Hakone, Japan.

[a] For data references, see W. J. Lackey, D. P. Stinton, G. A. Cerny, L. L. Fehrenbacher, and A. C. Schaffhauser, *Ceramic Coatings for Heat Engine Materials: Status and Future Needs*, ORNL/TM-8959, Oak Ridge National Laboratory, Oak Ridge, Tennessee.
[b] N, normal; T, total hemispherical.
[c] Calculated using $R = k\sigma(1 - \mu)/E\alpha$ (as R increases, thermal shock resistance increases).
[d] Corning grade 9606.
[e] Kennametal grade K-701.
[f] Scleroscope.
[g] Gray cast iron.

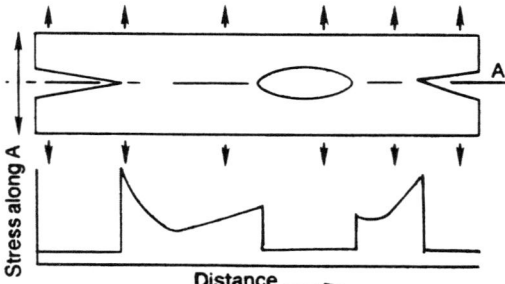

Figure 6.1 Stress distribution in a tensile plate with cutouts.

cracks can be related to the K_{1c}. Figure 6.2(a) illustrates such a test. Figure 6.2(b) shows a sample after such a test. Based on the average crack length, the indentor dimensions, and various properties of the ceramic, the fracture toughness is calculated from the following formula*:

$$K_{1c} \phi = 0.48(C/a)^{-3/2} H a^{-1/2}$$

where

$H = 0.47P/a^2$ (from Vickers test)
$P = $ load
$a = $ ½ indent diagonal
$C = $ crack length
$\phi = $ constraint factor $\simeq 3$

Figure 6.3 shows a plot of data obtained from such indentation tests on a variety of polycrystals and single crystals.

A second method of measurement is to prepare a standardized beam and insert a known defect in the form of a specified chevron-shaped notch. The specimen configuration is depicted in Figure 6.4. The specimen is then subjected to a four-point bending test. The fracture toughness is then calculated from the force required to fracture the specimen, its dimensions, modulus of elasticity, and Poisson's ratio for the material.

A third technique is the short-beam approach shown in Figure 6.5. Here a chevron notch is cut into a short beam loaded as indi-

---

*A. G. Evans and E. A. Charles, Fracture toughness determinations by indentation, *J. Am. Ceram. Soc.* **57**(7-8).

FRACTURE TOUGHNESS

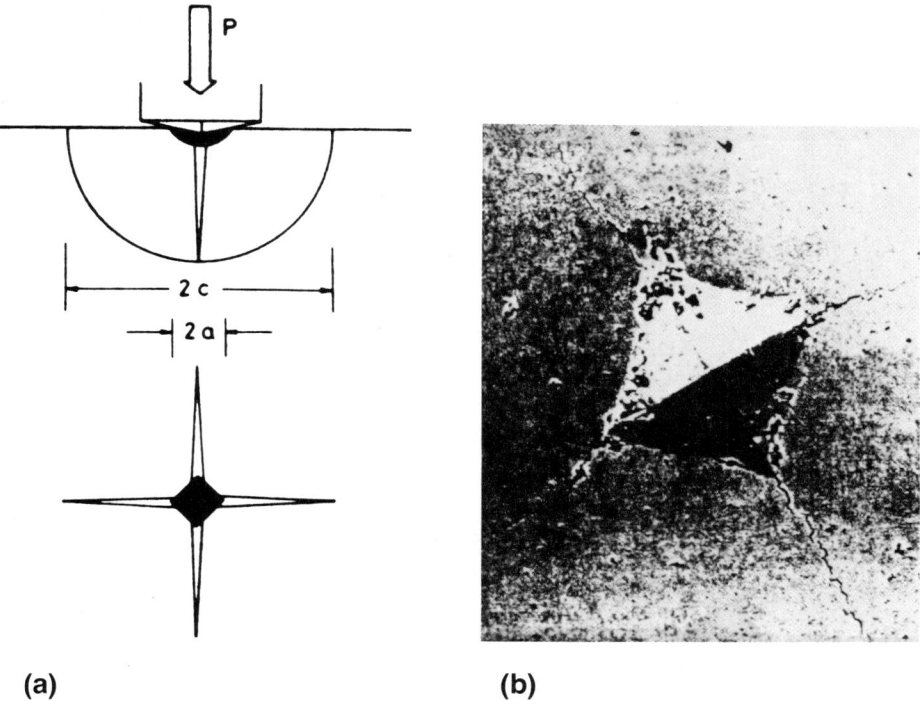

Figure 6.2 (a) Schematic of Vickers-produced indentation-fracture system, peak load P, showing characteristic dimensions c and a of penny-like radial/median crack and hardness impression, respectively. (b) Indentation crack in specimen.

cated in Figure 6.5. Again, the fracture toughness is calculated from the load required to fail the specimen, its dimensions, the modulus of elasticity, and Poisson's ratio for the material. However, it has to be realized that on a statistical basis small samples have greater strength than large samples, since the larger sample has a higher probability of containing a large flaw.

Single Crystals

Single crystals form the building blocks for poycrystalline ceramics. The fracture toughness of the single crystal is a critical parameter

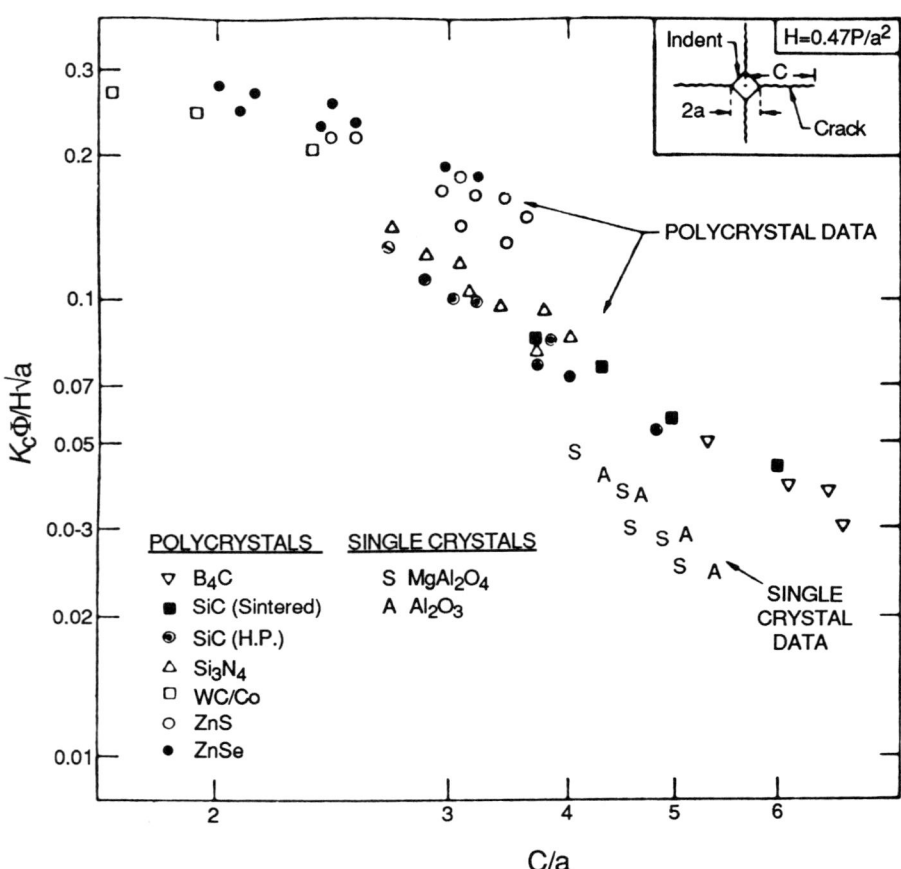

Figure 6.3 Radial crack extension plotted using normalized coordinates $K_c\Phi/H\sqrt{a}$ and C/a.

FRACTURE TOUGHNESS

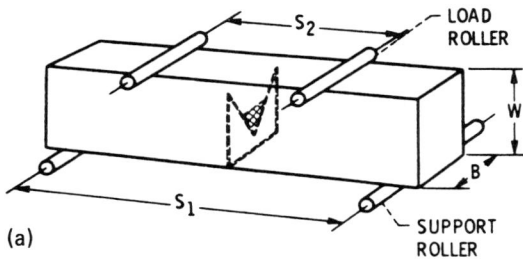

(a)

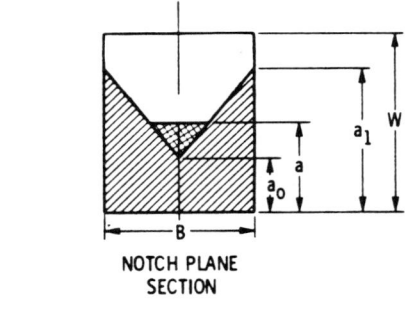

(b)

Figure 6.4 Four-point bend specimen with chevron notch.

governing the fracture toughness of the polycrystalline solid. Table 6.2 shows fracture toughness for a variety of single crystals. The K tends to increase with increasing elastic modulus. Mathematical models have been developed that enable accurate prediction of the fracture toughness of covalently bonded ceramics such as ZnSe and GaAs. Covalently bonded compounds are those where the atoms share electrons. Ionically bonded compounds, such as common salt (NaCl), are those where the atoms are totally ionized, each with its own set of electrons. The covalent structure of the diamond crystal is represented in Figure 6.6. This covalency accounts in part for the extremely high hardness of diamond. In fact, diamond is the hardest substance known.

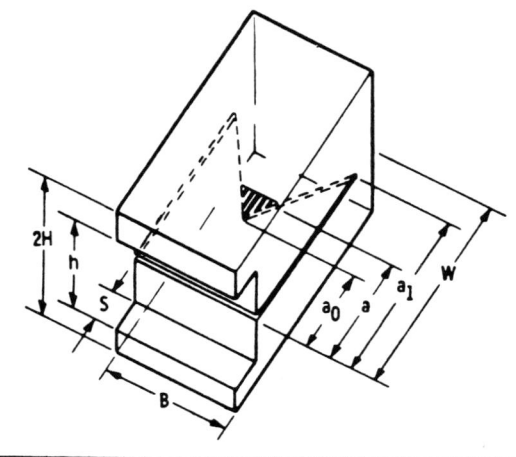

MATERIAL	B	W	H	W/H	h	S	a_0	a_1
Si_3N_4	8.8 - 9.6	15.2	4.5	3.4	6.3	2.5	2.5 - 7.3	15.2
Al_2O_3	25.4	50.8	12.7	4.0	12.7	3.8	10.6 - 22.2	50.8
	25.4	38.1	12.7	3.0	12.7	3.8	8.6 - 17.6	38.1
	12.7	25.4	6.35	4.0	6.3	3.8	4.9 - 11.5	10.2 - 25.4
	12.7	19.1	6.35	3.0	6.3	3.8	1.7 - 6.9	19.1

All dimensions in mm.

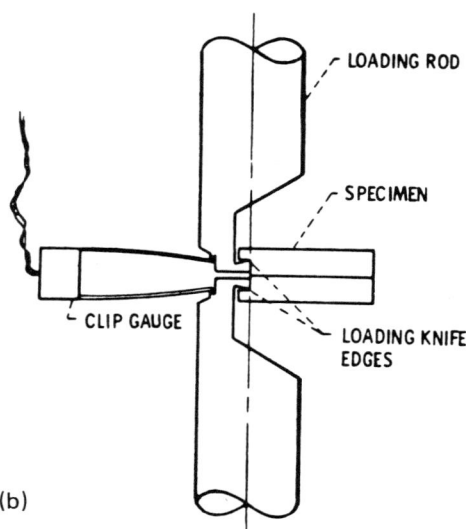

Figure 6.5 (a) Chevron-notch short-bar fracture toughness test specimen; (b) Setup for Si_3N_4 and 12.7-mm-thick Al_2O_3 short-bar specimen tests.

FRACTURE TOUGHNESS

Table 6.2 Fracture Toughnesses of Single Crystals

Material	K_{Ic} (MPa·m$^{1/2}$)
Al_2O_3	2.0
Si	0.8
GaAs	0.5
ZnSe	0.3
$BaTiO_3$	0.7
CaF_2	0.3
$MgAl_2O_4$	0.8
Y_2O_3	0.6

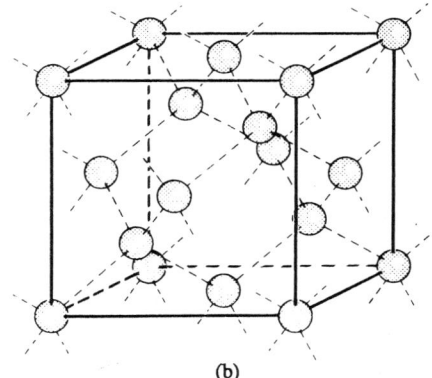

Figure 6.6 Diamond structure. The strength of the covalent bonds is what accounts for the great hardness of diamond. (a) Two-dimensional representation; (b) three-dimensional representation.

Figure 6.7 schematically illustrates the influence of the flaw size-to-grain size ratio on the fracture toughness of a polycrystalline ceramic for a given flaw size. As the grain size becomes smaller relative to the flaw size, the fracture toughness tends to increase.

TOUGHENING OF CERAMICS

The effort to increase the fracture toughness of ceramics has taken a position in the forefront of research efforts around the world. As mentioned in an earlier chapter, the driving force is to employ

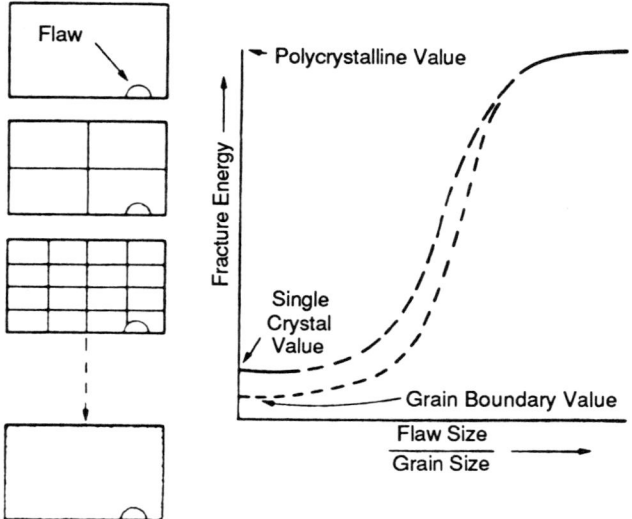

Figure 6.7 Schematic of flaw size/grain size effects on fracture energy.

ceramics in heat engines at higher temperatures than is possible with metals and thus improve the Carnot efficiency of these engines with extremely high payoff in terms of reduced fuel expenditures.

The fracture toughness of ceramics can be enhanced by a variety of mechanisms. Microstructure can be engineered to cause any crack to have an increased difficulty in propagating by requiring the crack to advance through a tortuous path. A variety of mechanisms can be employed to accomplish this objective. When the crack propagation path is interrupted due to misorientation of the easy crack propagation direction that occurs at a grain boundary, the energy required for the crack to propagate is increased. If one looks at the fracture surface, the roughness of that surface gives an indication of the toughness of the material. For a given grain size, the surface roughness is dependent on the misorientation of the grains. The crack plane in a single crystal is normally quite smooth.

Microcracking of the structure is another mechanism that leads to increased fracture toughness. For noncubic crystals, where signif-

FRACTURE TOUGHNESS

icant anisotropy exists in the thermal expansion, stresses induced during cool-down from processing temperatures cause microcracks. For alumina (Al_2O_3) it has been demonstrated empirically that the fracture toughness reaches a maximum for grain size of 60 μm, as shown in Figure 6.8. Toughening due to microcracking does not necessarily lead to strengthening, since the microcracks link and form enlarged flaws.

Crack growth occurs at ambient temperatures in almost all ceramic bodies, due to the effects of the environment, usually water, although other substances can be active. Crack growth rates can

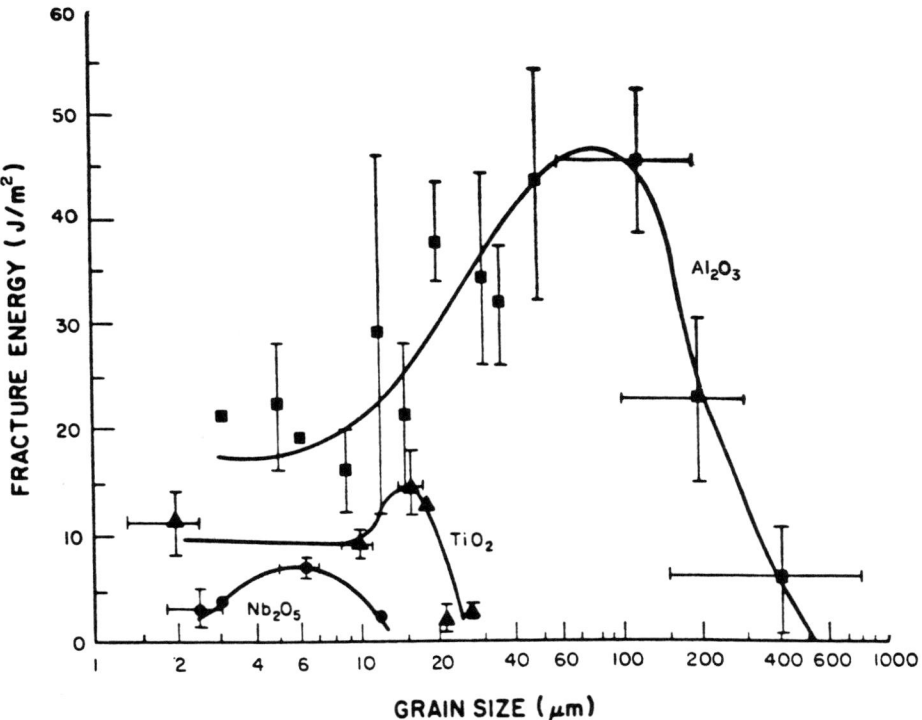

Figure 6.8 Fracture energy of Al_2O_3, TiO_2, and Nb_2O_5 as a function of grain size. All of these materials have a noncubic crystal structure. Data come from a number of investigators.

vary from 10^{-3} to 10^{-11} m/s. Crack growth rates are dependent on the relative humidity of the active substance in which the crack is growing.

Silicon dioxide is sintered with 8 to 15 wt % of a sintering aid which allows the body to consolidate by pressureless sintering. The crack encounters odd-shaped grains and grain boundaries rich in the sintering aid. The crack path must constantly change its direction, and thus increased energy is required for the crack to grow, and consequently the toughness of the ceramic is increased. In cemented tungsten carbides, the sintering aid is a metal capable of plastic deformation. Thus the crack is blunted and much more energy is consumed during crack propagation, resulting in increased fracture toughness. Figures 6.9 and 6.10 show typical microstructures of silicon nitride and cemented tungsten carbide.

Very fine, relatively short fibers called *whiskers* can be incorporated into the ceramic body in volume fractions up to 25%. The presence of these whiskers causes cracks to be impeded in their propagation. Energy is absorbed by whisker pullout from the matrix, whisker bending, and fracture. However, incorporation of the whiskers in the ceramic body poses difficult problems. To be effective the whiskers must be stiffer (i.e., have higher Young's modulus) than the matrix, of approximately the same thermal expansion coefficient, and chemically compatible at the high temperatures required for consolidation.

The bond strength between the whisker and matrix must be carefully controlled. Too great a bond strength prevents pullout and consequent energy absorption. On the other hand, too little bond strength does not produce very much energy absorbtion during pullout and therefore little enhancement in fracture toughness. Bond strength can be controlled by compositional adjustments to the sintering aids or by special coating on the whiskers. Figure 6.11 is a schematic of whisker pullout illustrating how the fibers pull out of the matrix and then fracture as the crack develops, absorbing energy.

A successful application of whisker toughening is in hot-pressed whisker-reinforced alumina for metal-cutting applications. The whiskers are silicon carbide (SiC). The whiskers increase the fracture toughness of unreinforced aluminum oxide by a factor of 2.

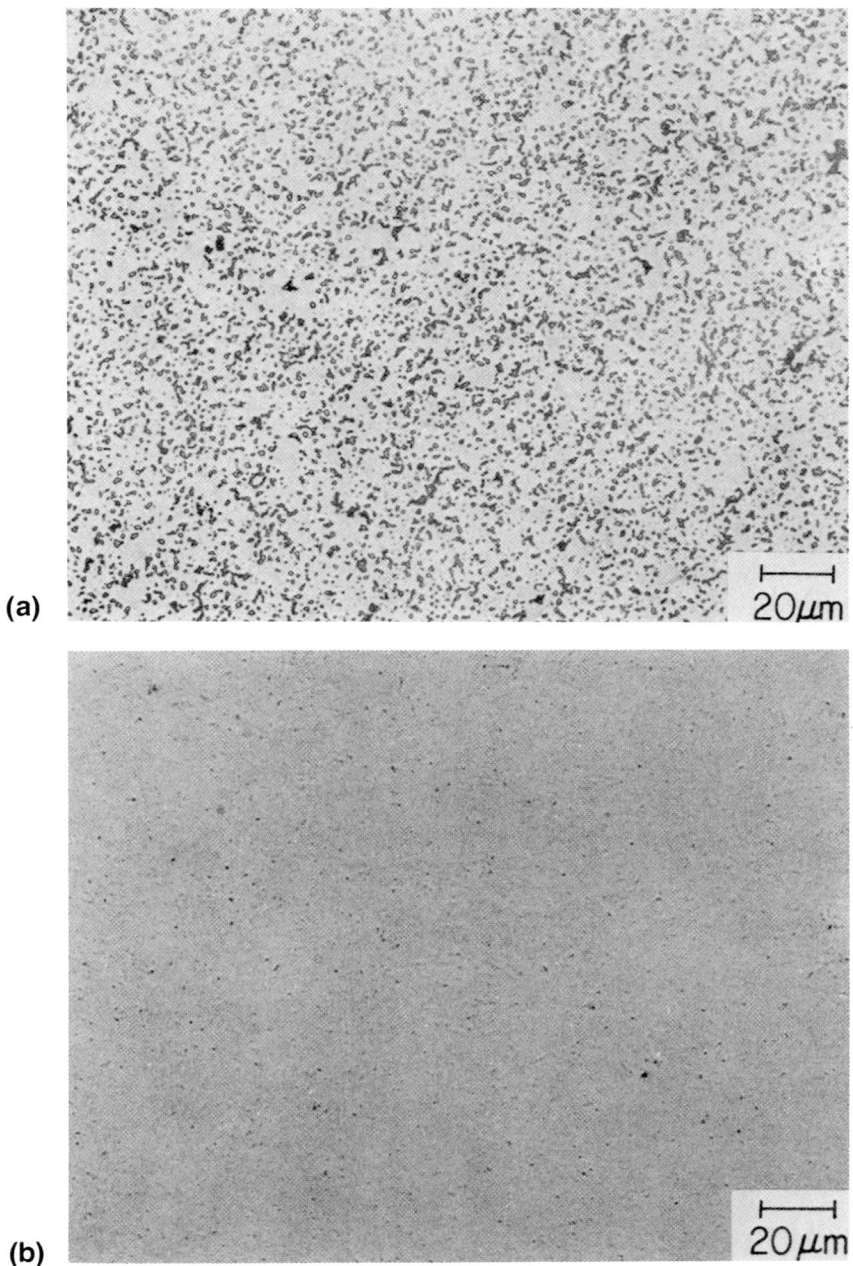

Figure 6.9 Residual porosity in compacts of Si_3N_4 + 7% $BeSiN_2$ + 7% SiO_2 sintered (a) in 2.1 MPa of N_2 and (b) by the two-step, N_2 pressurizing method.

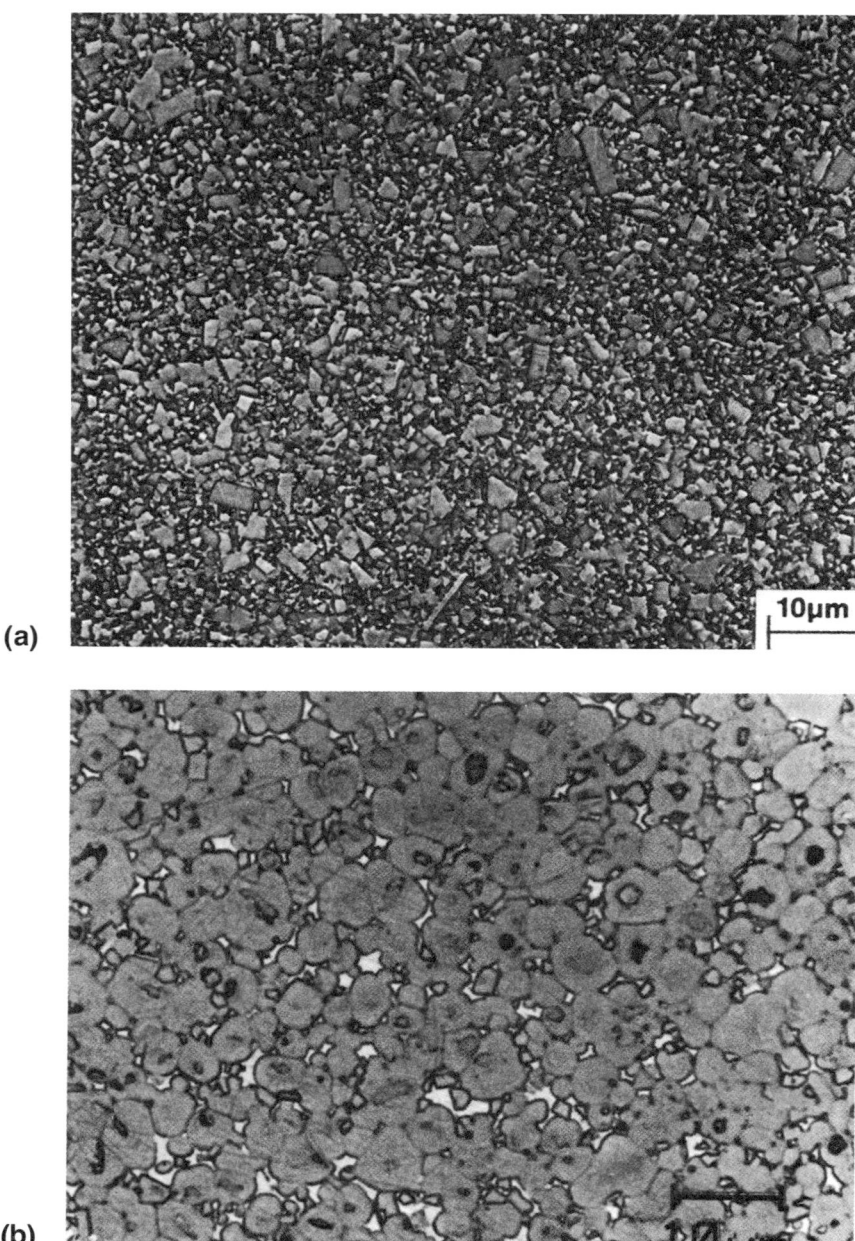

Figure 6.10 Micrographs of two cutting tools grades of cemented tungsten carbide: (a) a straight WC grade in a cobalt binder showing angular WC morphology; (b) a complex multicarbide (WC-TiC-TaC) grade in a cobalt binder showing round multicarbide morphology.

FRACTURE TOUGHNESS 117

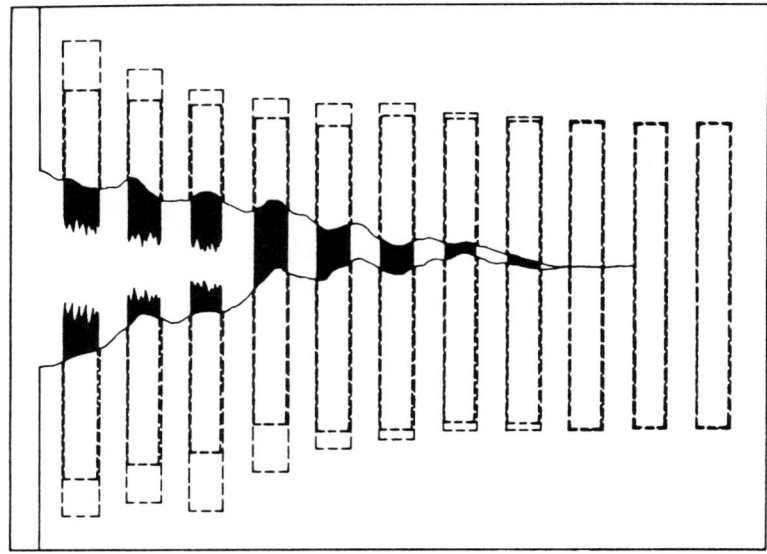

Figure 6.11 Whisker pullout absorbs energy during propagation of crack in whisker-reinforced ceramic-matrix composite. Pullout is believed to be the principal toughening mechanism in this type of composite.

The composite is used for rough machining of superalloys. The material is not suitable for ferrous alloy cutting because the SiC reacts with iron at the cutting temperatures.

The variety of fibers that may be used to reinforce ceramics is broadening. Table 6.3 lists commercially available fibers and whiskers and some of their properties. Table 6.4 shows the recently developed ceramic fibers and their properties. Some of these fibers are not available except in experimental quantities. The key point is that significant technical effort is being expended to develop superior fibers. All of these materials are susceptible to degradation when subjected to elevated temperatures over long periods of time. The most widely used of these fibers is Nicalon, a SiC fiber containing some oxygen and excess carbon. These components are a blessing and a curse since they facilitate bonding between the matrix and the fiber but also lead to chemical instabilities at elevated temperature for extended periods.

Table 6.3 Ceramic Fibers That Have Been Available for Some Time[a]

Manufacturer	Designation	Composition (wt %)	Tensile strength (MPa)	Tensile modulus (GPa)	Density (g/cm³)	Diameter (μm)
Nippon Carbon	Nicalon	59 Si, 31 C, 10 O	2520-3290	182-210	2.55	10-20
Avco	SCS-6	SiC on carbon core	3920	406	3.0	143
3M	Nextel 312	62 Al_2O_3, 14 B_2O_3, 24 SiO_2	1750	154	2.7	11
Du Pont	FP	>99 α-Al_2O_3	>1400	385	3.9	20
Sumitomo		85 Al_2O_3, 15 SiO_2	1800-2600	210-250	3.2	9-17

[a]All data were provided by fiber manufacturers and their distributors. Mechanical property data are for room temperature. Fiber gage lengths tested may differ, making direct comparison of tensile strength data questionable.
Source: Ceram. Bull. **66**(2), Feb. (1987); copyright American Ceramic Society.

Table 6.4 New Ceramic Fibers[a]

Manufacturer	Designation	Composition (wt %)	Tensile strength (MPa)	Tensile modulus (GPa)	Density (g/cm³)	Diameter (μm)
Ube	Tyranno	Si, Ti, C, O	>2970	>200	2.3-2.5	8-10
Avco		Si, C	>2800	280-315	2.3	6-10
Dow Corning/Celanese	MPDZ	47 Si, 30 C, 15 N, 8 O	1750-2100	175-210	2.3	10-15
Dow Corning/Celanese	HPZ	59 Si, 10 C, 28 N, 3 O	2100-2450	140-175	2.35	10
Dow Corning/Celanese	MPS	69 Si, 30 C, 1 O	1050-1400	175-210	2.6-2.7	10-15
3M	Nextel 440	70 Al_2O_3, 28 SiO_2, 2 B_2O_3	2100	189	3.05	10-12
3M	Nextel 480	70 Al_2O_3, 28 SiO_2, 2 B_2O_3	2275	224	3.05	10-12
Du Pont	PRD-166	Al_2O_3, 15-25 ZrO_2	2100-2450	385	4.2	20

[a]All data were provided by fiber manufacturers and their distributors. Mechanical property data are for room temperature. Fiber gage lengths tested may differ, making direct comparison of tensile strength data questionable.

FRACTURE TOUGHNESS

Yet another means of toughening is by phase transformation. This method has been demonstrated in toughening of zirconia (ZrO_2). In phase-transformation-toughened zirconia, small grains of a metastable second phase are incorporated by appropriate compositional and thermal control in a matrix of a thermodynamically stable first phase such that both phases are in metastable equilibrium with each other at room temperature.

The phase diagram for the ZrO_2-Y_2O_3 system is shown in Figure 6.12. For this type of toughening, a composition of approximately 2.0 mol % (3.6 wt %) Y_2O_3 is used. The material is brought to equilibrium in the tetragonal phase at a temperature of about 1100 °C. The body is then cooled at a relatively quick rate. Since the thermal transformation of the tetragonal phase to the monoclinic phase is sluggish, the ZrO_2 retains a good fraction of the tetragonal phase at room temperature in a matrix of monoclinic ZrO_2.

The difference between the lattice of the tetragonal and monoclinic phases is shown in Figure 6.13. It is the transformation of the metastable tetragonal phase to the thermally stable monoclinic phase under an applied stress that toughens the ZrO_2 body. During this transformation, the transforming crystallites experience an increase in volume due to the atomic reorientation. The volume change has to be high enough to create a stress state around the transforming particle while not inducing so high a stress as to create a failure. The induced stress state impedes the advancing crack. The volume change is a consequence of the particular system and is a fact of nature.

Transformation toughening has been demonstrated in predominantly zirconia ceramics and also in other types of matrixes. Figure 6.14 illustrates transformation toughening around an advancing crack tip where tetragonal zirconia-hafnia solid solution has been dispersed in a mullite matrix. The stress field developing around the advancing crack tip causes the tetragonal particles to transform to the monoclinic structure, absorbing energy and impeding the crack growth. The microstructure shown in Figure 6.14 illustrates the second-phase particles (white) in a matrix of mullite. In transformation-toughened zirconia, fourfold increases in fracture toughness are achieved at room temperature. Unfortunately,

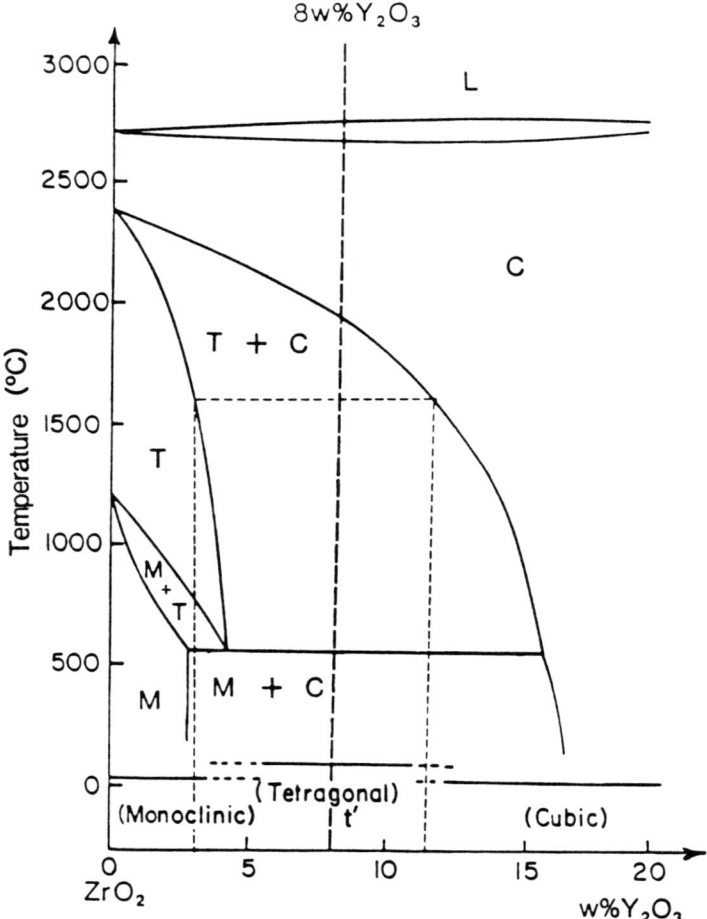

Figure 6.12 Phase diagram for the system ZrO_2-Y_2O_3.

FRACTURE TOUGHNESS

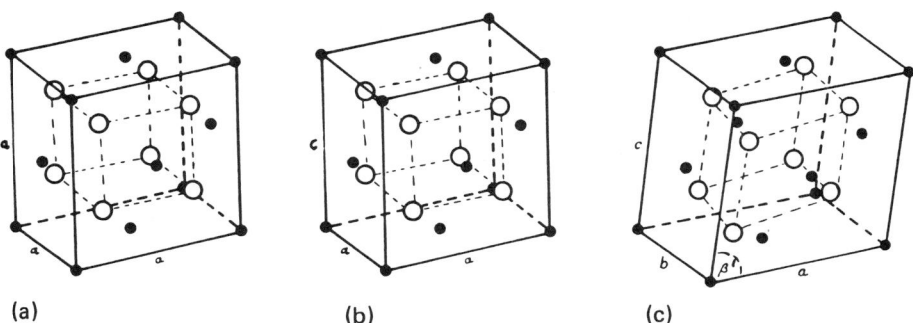

Figure 6.13 Three polymorphs of ZrO_2: (a) cubic phase, (b) tetragonal phase $c/a \approx 1.02$, (c) monotinic phase. [From N. Claussen, M. Rühle, and A. H. Heuer (ed.), *Science and Technology of ZrO_2*, Vol. II, American Ceramic Society, 1984, Fig. 1, p. 3.]

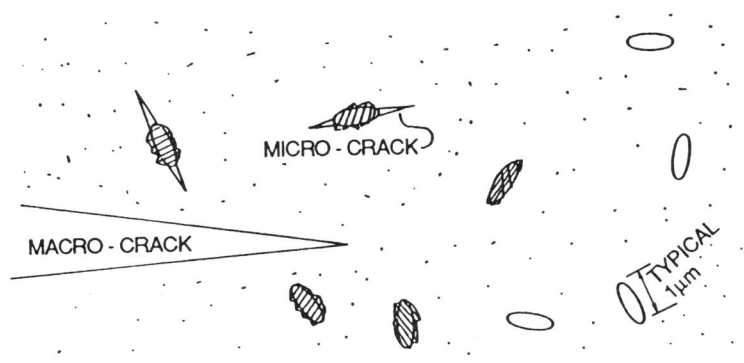

	ZrO_2 Transforms in Stress Field of Crack Tip, Complex Stress and Strain Field Impedes Crack Propagation	
Candidate Matrixes	Al_2O_3	Tetragonal → Monoclinic, $+ \triangle V$
	$3 Al_2O_3 \cdot 2 SiO_2$ (MULLITE)	*Toughener* ZrO_2/HfO_2
	SiALON	Solid Solution

Figure 6.14 Schematic of transformation toughening. (From S. Musikant, H. Rauch, and E. Feingold, "Transformation Toughening Ceramics for Engines," *Transactions Automotive Contractors Coordination Meeting,* 1983, Dearborn, Michigan, p. 138.)

as the service temperature is increased, the toughening effect is reduced, so for high-temperature applications, the effect is not as dramatic.

Surface flaws on a ceramic component can be reduced by various manufacturing techniques. One class of treatments induces a compressive stress in the surface. In this concept, the tensile stresses needed to open up a crack cannot do so until the induced compressive stress is overcome, thus requiring greater force to effect a failure. These compressive stresses can be induced by "stuffing" the surface structure with atoms that are larger than the existing atoms, thus producing compressive stress in the surface. This stuffing can be accomplished chemically.

Another method of inducing surface stresses is by heating the body and then quenching in a controlled manner so that the core, which cools more slowly than the surface, pulls the surface in as the core contracts, inducing compressive stresses in the surface when the entire body reaches a uniform temperature. A similar effect can be generated by chemical surface treatment at elevated temperature where the larger atoms fit in the structure, but become crowded after cooling of the body, thus inducing compressive stresses in the surface. Normal surface flaws can be blunted simply by heating and allowing the flaws to heal or by surface coating using chemical vapor deposition methods or glazes. These types of surface treatments have been effective in alumina, silicon carbide, and silicon nitride, where the apparent strength of the materials has been doubled by such methods.

Glass tubes and containers can be strengthened by the application of a thin glaze by a flame-spraying technique. The glazed tubes are on the order of 2.75 times stronger than the unglazed variety. The glaze on container glass (soda-lime-silica) induces a compressive stress in the surface (wall thickness 3 mm) on the order of 200 MPa (29,000 psi), where the glaze is 50 μm thick, with an expansion coefficient of 40×10^{-7} C^{-1}. The container glass has a much higher coefficient of thermal expansion, 103×10^{-7} C^{-1}. After the glaze is applied, the containers are reannealed. That is, they are given a thermal treatment such that the desired compressive stress in the glaze is achieved.

7
Joining of Ceramics

In most designs, the ceramic component must be joined or fastened to another ceramic part or to a metal piece. Threads and threaded fasteners can be employed, but these are difficult and expensive to provide in ceramic bodies. Glass-to-metal seals are highly developed, such as the metal fitting in a light bulb. Joining oxide ceramics to each other, nonoxide ceramics to each other, and oxide to nonoxide ceramics present increasingly difficult procedures. Figure 7.1 shows the different categories of joining that are used. Except for mechanical fasteners, all the methods shown require elevated temperatures. Joining can take place by a liquid interface which reacts in some way with the end members, or by diffusion, either self-diffusion or heterogeneous diffusion. The control of the narrow interface between the two end members is the critical requirement for a successful joint. Properties may change rapidly at the interface. These changes may be:

1. Crystallographic: leading to a lattice mismatch
2. Electronic: bonding may go from covalent to ionic or metallic
3. Mechanical: variation in elastic modulus

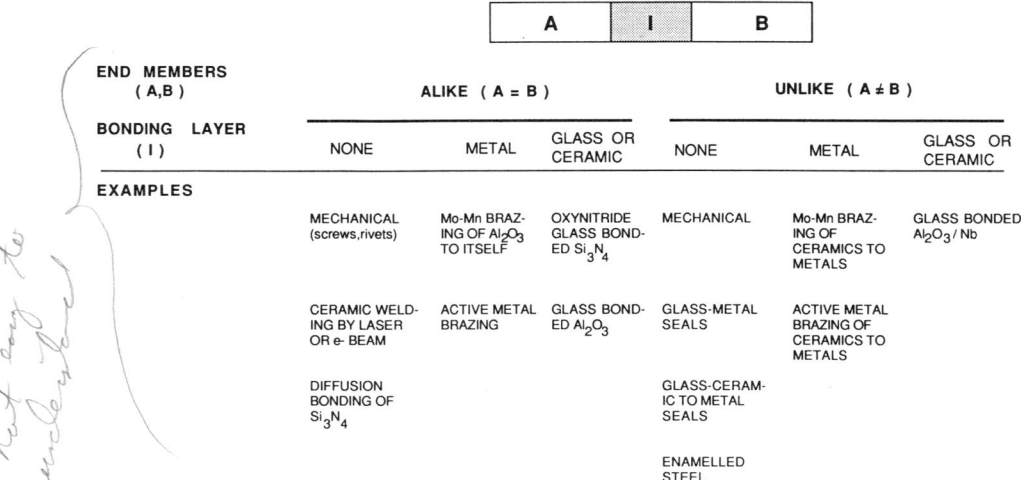

Figure 7.1 Different types of glass and ceramic joining.

4. Thermodynamic: leading to reactions at elevated temperature
5. Thermomechanical: differences in coefficient of thermal expansion leading to thermal stresses

For a joint to be successful, these discontinuities must be accomodated. Maximum strength requires the formation of a chemical bond at the interface. Coefficients of expansion need to be matched to reduce stress gradients through the joint. The formation of an intimate atomic contact interface during the bonding process is essential to the formation of the chemical bond. For a solid-to-solid contact, such as in a diffusion bond, pressure is needed. In liquid-phase joining the liquid must wet the contacting solid. The reaction must proceed thermodynamically to equilibrium.

REDOX REACTION

Glass-metal and oxide-ceramic/metal bonds involve so-called redox reactions where the metal is oxidized and the cation of the ceramic or glass is reduced. The metal-oxide formed saturates the interface, forms a layer, and ultimately chemically bonds to both parents since

JOINING OF CERAMICS

the metal-oxide is compatible with the metal and the ceramic. For metal joints to oxide ceramics or glass, the metal is preoxidized to create a favorable interface. The α-Al_2O_3/niobium seal has been studied extensively. The materials possess an excellent match of coefficient of thermal expansion, 81×10^{-7} C^{-1} for Al_2O_3 against 78×10^{-7} C^{-1} for Nb (niobium) between room temperature and 1000 °C. When two flat pieces of these two materials are placed in contact under pressure at about 1700 °C, strong bonding occurs. There develops a gradient of O and Al into the Nb. Both phases maintain their crystalline form.

A modified process would be the use of a shim of metal, such as Nb, between two pieces of ceramic, such as α-Al_2O_3. For other combinations of ceramics, appropriate metals have to be discovered which will form strong and stable chemical bonds to the ceramics. In the case of the Nb/α-Al_2O_3 bond, the 1700 °C temperature required is formidable from a processing standpoint. Therefore, the use of a glass layer between the two end members has been developed to permit the bonding to take place at considerably lower temperature of 1400 °C.

SOLDERS

One can also use compatible solders (melting point in the range of 400 °C) or brazes (melting temperature in the range of 900 °C). The use temperature, of course, has to be lower than the processing temperature. A braze that is widely used is Cu · xAg, where x can be varied to adjust the properties of the braze. However, this braze material does not wet or chemically react with the ceramic. Therefore, a small percentage of a reactive metal such as titanium is added to the braze composition. In the case of Al_2O_3, the Ti undergoes a redox (reduction-oxidation) reaction that causes wetting of the ceramic surface with the braze liquid and the formation of an oxide at the interface that forms the chemical bond with bond end members.

This is illustrated in Figure 7.2(a). A similar process is shown in Figure 7.2(b) (left-hand side). Here the Ti is applied to the ceramic surface by painting a slurry of titanium hydride followed by drying and then firing in an atmosphere and temperature where the hydride

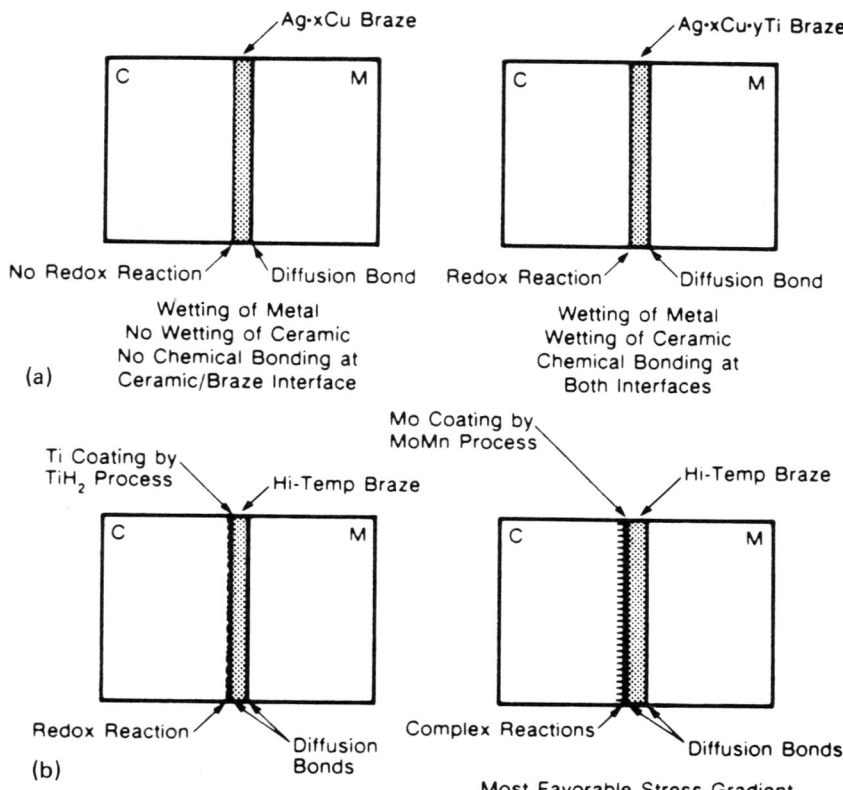

Figure 7.2 Schematic illustrating several conditions under which brazes are utilized: (a) braze in contact with unmetallized ceramic; (b) braze in contact with metallized ceramic.

decomposes, leaving Ti on the ceramic surface which undergoes a redox reaction with a residue of Ti oxide on the ceramic surface. The Ti oxide then forms the chemical bond with the braze metal.

GLASSY BOND

Figure 7.2(b) illustrates another type of process where the assembly is effected by the formation of a glassy phase that bonds to the metallizing as well as to the ceramic grains. In this case, a paint of Mo and Mn or their oxides is applied to the ceramic and fired in a hydrogen atmosphere furnace with a controlled H_2O content. This environment causes the MnO to react with the ceramic grains, while Mo metal sinters as a porous coating into which the MnO glassy phase infiltrates. The glass reacts with the Mo, forming a chemical, hence strong, bond. The Mo coating is then Ni-plated, which provides a clean and continuous surface to which the braze metal bonds in the next operation. Each specific metal-glass-ceramic joining combination requires process development to define the exact compositions and procedures to lead to a successful joint.

NONOXIDE CERAMICS

Joining of nonoxide ceramics is less highly developed than that for oxide ceramics. The nitrides and carbides present a different chemical problem. The nonoxide ceramics are mainly joined by metal fillers and braze alloys or reactive glasses or ceramics. Most nonoxide ceramics have significantly lower coefficients of expansion than the metals to which they are to be joined. Therefore, ductile metals are used as the brazing medium to allow stress redistribution as the joint cools down from the brazing temperature. Since most of the nonoxide ceramics incorporate some oxides in the form of sintering aids, glass or ceramic bonds with compositions close to that of the glassy phase in the ceramic being joined can often be developed.

Nonoxide ceramics have been joined by reactive metals. Ti or Al can be used to join Si_3N_4 or AlN. Pressureless sintered SiC ceramic bodies have been joined successfully using a mixture of sinterable submicrometer α-SiC powder containing 6% Al, 1% B, and 1% C placed between the two bodies being joined. In this work the joint was hot pressed at 1650 °C for 0.5 hour under a pressure of 50 MN/m^2. The resulting bond strength of 400 MN/m^2 was as strong as the sintered body itself.

8
Nondestructive Testing

Since flaw population and flaw morphology are critical to successful application of high-performance ceramics, the development and application of nondestructive test (NDT) techniques are essential. Flaws are detected in the green state and in the fired state. If flaws are detected in the green body, the expense of processing a part that is no good is avoided.

RADIOGRAPHY

Conventional radiography uses energetic gamma-ray or x-ray sources and very fine grain film to detect flaws in solid bodies. A simple diagram of the principle of radiographic inspection is shown in Figure 8.1. The size of the defect that can be detected in this manner is dependent on the thickness of the part, its x-ray absorption characteristics, the size of the flaw, the orientation of the flaw, and the x-ray opacity of the flaw relative to the part. The x-ray image has to be taken in more than one orientation to observe all the defects since thin defects such as closed cracks will not be observed when the x-rays are parallel to the crack.

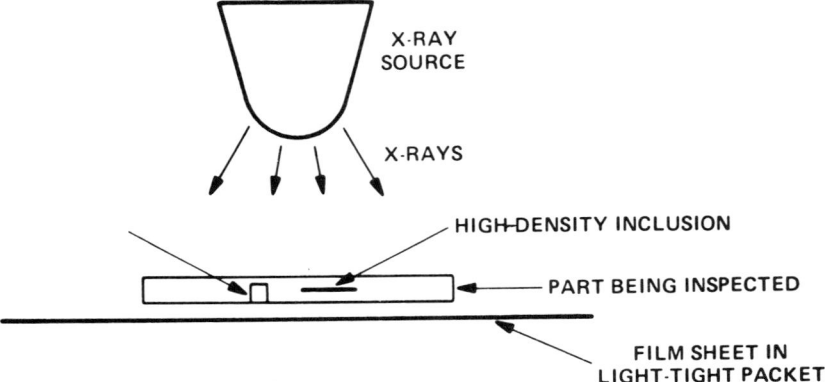

Figure 8.1 Schematic of conventional x-ray radiography setup.

Ceramics of low-atomic-number elements such as Si, Al, C, N, and O are relatively transparent to x-ray, while W, Fe, and other dense metals are relatively opaque. The resolution capability is expressed as a percentage of the part thickness; thus inspection at the $2T$ level means that the method must be able to detect flaws 2% of the thickness of the part or larger. Microfocus x-ray employs a special x-ray tube that focuses the x-ray beam to a spot 0.05 mm (0.002 in.) in diameter. This is especially useful in inspecting very small critical regions, such as the trailing edge of a ceramic rotor or stator blade for gas turbines.

Image enhancement techniques are also available. Here the photographic plate is backlighted and picked up on a video camera. The image is then digitized into a pixel array of 480 × 512 pixels. The intensity of each pixel is assigned a gray value ranging from 0 (black) to 255 (white). These data can be operated on by computer programs to further enhance the image. The enhanced image is then displayed on a video tube and can be photographed for a permanent record. Reading of this image as of the original radiograph requires skill and experience to interpret the information appropriately. A schematic diagram of the image enhancement technique is shown in Figure 8.2. An enhanced image of graphite inclusions in hot-

NONDESTRUCTIVE TESTING 131

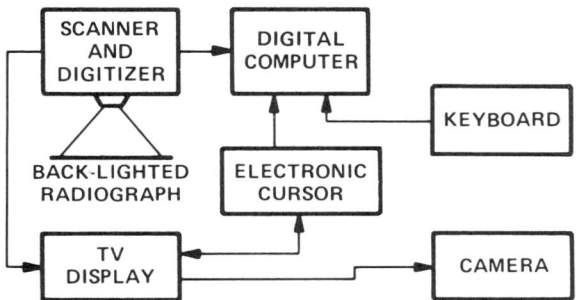

Figure 8.2 Schematic of the image enhancement system.

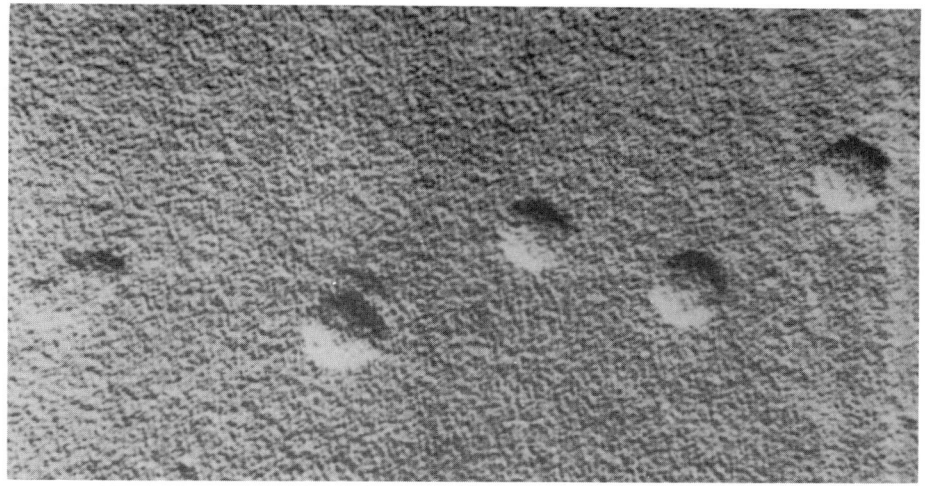

(a)

Figure 8.3 Image enhancement of 500-μm (0.02-in.) graphite inclusions in hot-pressed Si_3N_4. (Courtesy of Garrett Turbine Engine Company, Phoenix, Ariz., Division of the Garrett Corporation.)

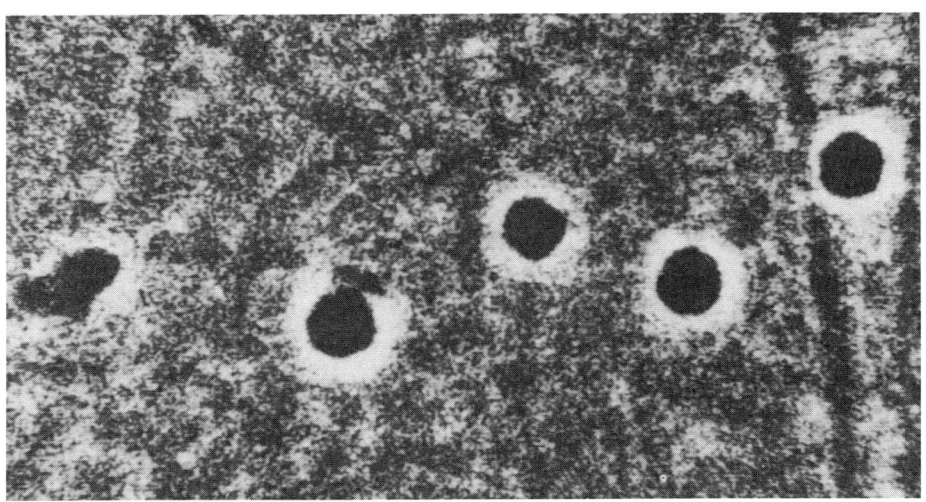

(b)

Figure 8.3 (continued)

pressed Si_3N_4 is shown in Figure 8.3(b) compared to the original image shown in Figure 8.3(a).

X-ray computed tomography (CT) is being used for intricate engine parts. This technique provides sections of the part in any aspect so that a complete portrayal of the part is available. X-ray CT can determine density gradients in green or fired parts with a high degree of sensitivity. Whereas conventional radiography is good to 1 to 2% variation in density, x-ray CT is about 100 times more sensitive. However, the machines for doing this type of work are high-cost assets.

ULTRASONIC NDE

Ultrasonic NDE (nondestructive evaluation) can be used to detect subsurface flaws in ceramics. A schematic showing the principle of this method is given in Figure 8.4. A piezoelectric transducer near the part emits ultrasonic waves that pass through the part. Whenever these waves hit a discontinuity in the material, secondary waves are generated due to scattering or reflection. A receiver detects these

NONDESTRUCTIVE TESTING

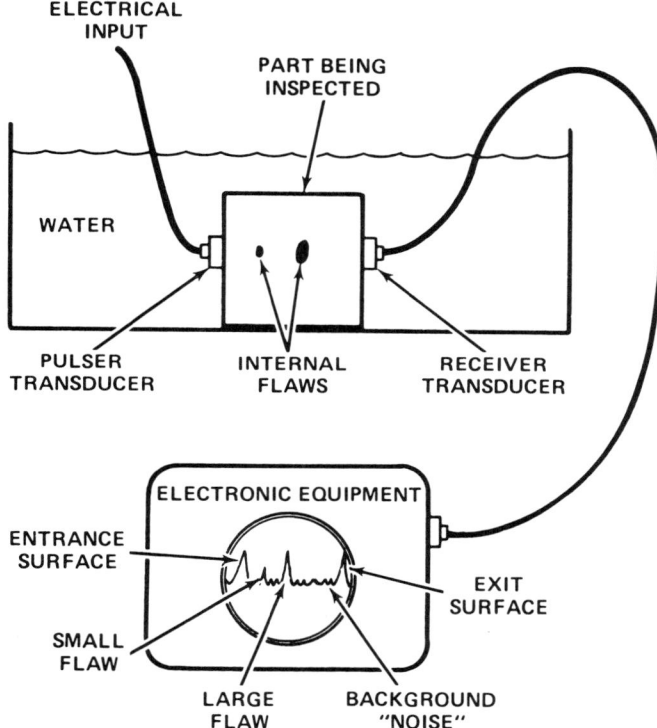

Figure 8.4 Schematic illustrating the basic principles of ultrasonic NDI.

secondary waves and converts them to an image. As in the case of radiography, the inspection has to be made in more than one orientation. Computer-aided ultrasonic inspection is being developed which greatly improves resolution. Ultrasonic inspection is best for flat-sided parts while complex shapes present patterns that are difficult to interpret.

DYE PENETRANT

Penetrant fluids are employed to detect surface flaws. In this method a fluorescent dye is painted on the part and then dried. Wherever there is a surface crack or pore, the penetrant is trapped and observed by means of an ultraviolet light. However, if the part has interconnecting porosity, the method will not work since the entire part will fluoresce under UV inspection.

MAGNETIC RESONANCE IMAGING

Magnetic resonance imaging (MRI) is a technique for imaging the body in three dimensions to provide planar representations. It operates by switching of magnetic fields and pulsed radio-frequency (RF) transceiver coils. The imaging is sensitive to density and chemistry and thus can be used to study the detailed chemistry of the body. This technique can be applied to injection molding processes to assess the homogeneity of the various organics (binder, plasti-

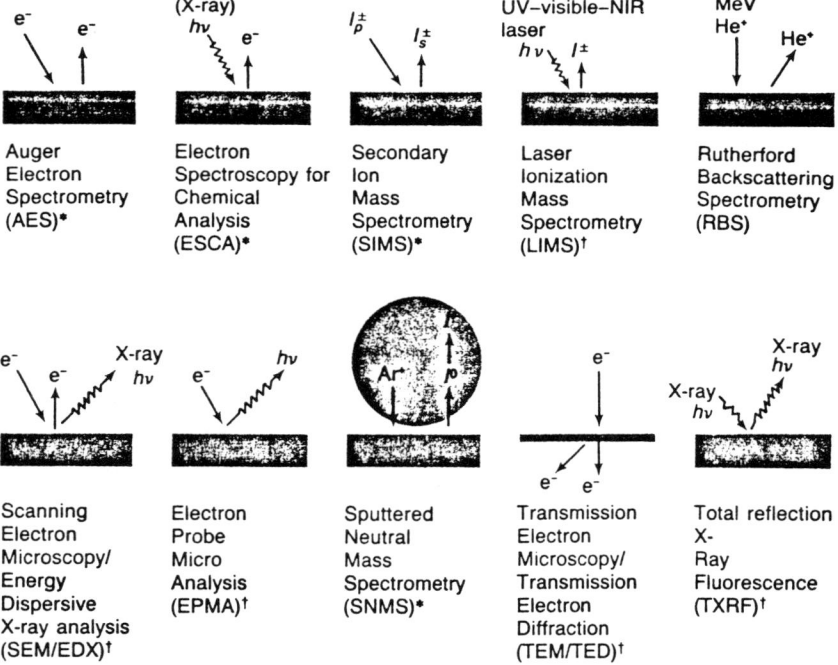

Figure 8.5 Surface analysis techniques.

NONDESTRUCTIVE TESTING

cizer, and mold release agents) used as the carrier for the ceramic powder. Such inhomogeneities affect subsequent processing and the mechanical properties of the part. MRI is a very expensive tool and is most likely to be employed in a research setting.

SURFACE ANALYSIS

The chemistry of surfaces can be characterized by a host of techniques. In these techniques, a primary radiation (probe beam) excites the surface, which emits a response. The response is observed as a spectrum that can be detected, recorded, and identified. Ten such techniques are illustrated in Figure 8.5. In these figures the following symbols are used: e^-, electron; $h\mu$, photon, frequency μ; and I, ion, positive ($^+$), negative ($^-$), neutral (^{0}), primary ($_p$), secondary ($_s$), argon ion (Ar^+). These spectroscopic methods are essentially nondestructive as long as the specimen can be accommodated in the instrument. In the case of transmission electron microscopy or diffraction, a special thin sample has to be prepared. As a practical matter, most objects are too big to fit into the examination chamber and small representative samples have to be used.

9
Ceramic Cutting Tools

The drive to increase productivity in the metal finishing industries worldwide has placed increasing demands on cutting tools. As the speed of metal removal increased, the need for more refractory materials for these cutting tools became evident and a consequent development of ceramic cutting tools ensued. High-speed steels, cemented carbide tool tips, coated carbide cutting tools, diamond and cubic boron nitride (Borazon*) tools, and recently silicon nitride and various alumina composites were successively developed. Figure 9.1 illustrates the increase in cutting speeds during the twentieth century.

TOOL WEAR

Severe stresses and temperatures occur at the cutting edge of the tool illustrated in Figure 9.2. The tip temperature may be as high as 1100 °C or higher, depending on the material of the tool bit as well as that of the work, speed, rate of feed, and lubricants and coolants

---

*Trademark of General Electric Co.

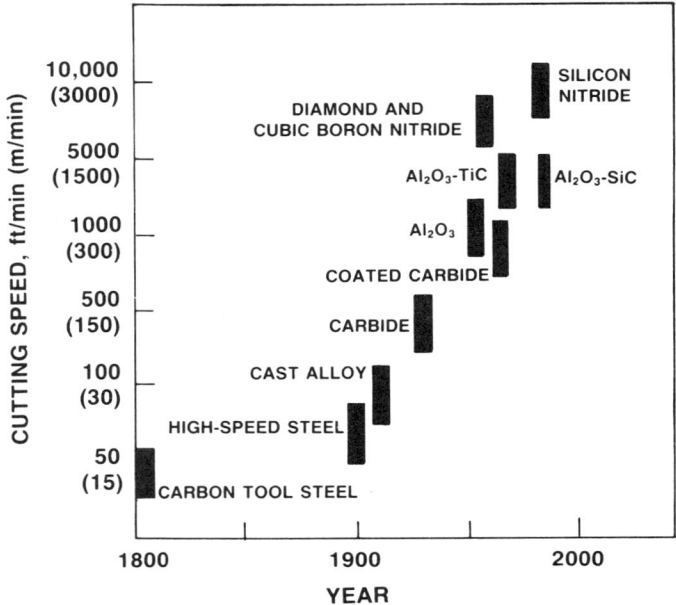

Figure 9.1 Change in productivity due to the introduction of new cutting tool materials.

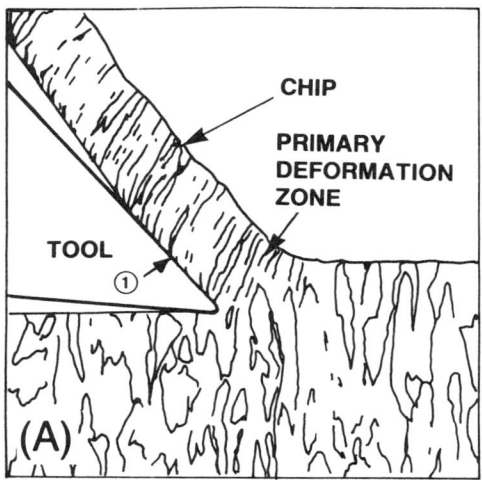

Figure 9.2 (A) Representation of the chip-forming process in metal cutting. (B) The temperature distribution near the tool surface. (C) Representation of tool wear. [From *Ceram. Bull.* **67**(2), Feb. 1988; copyright American Ceramic Society.]

CERAMIC CUTTING TOOLS

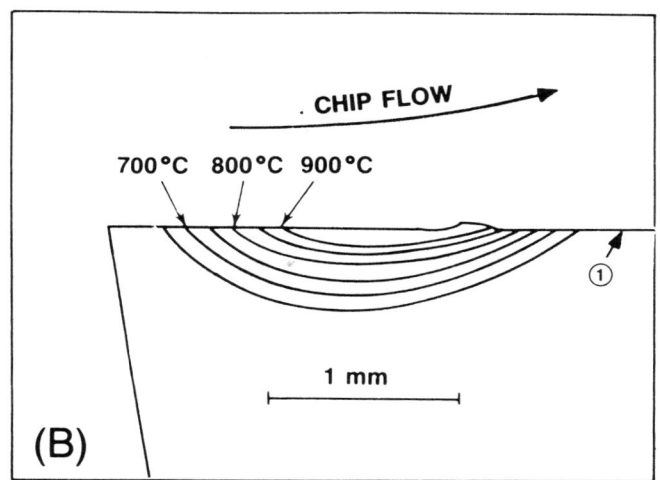

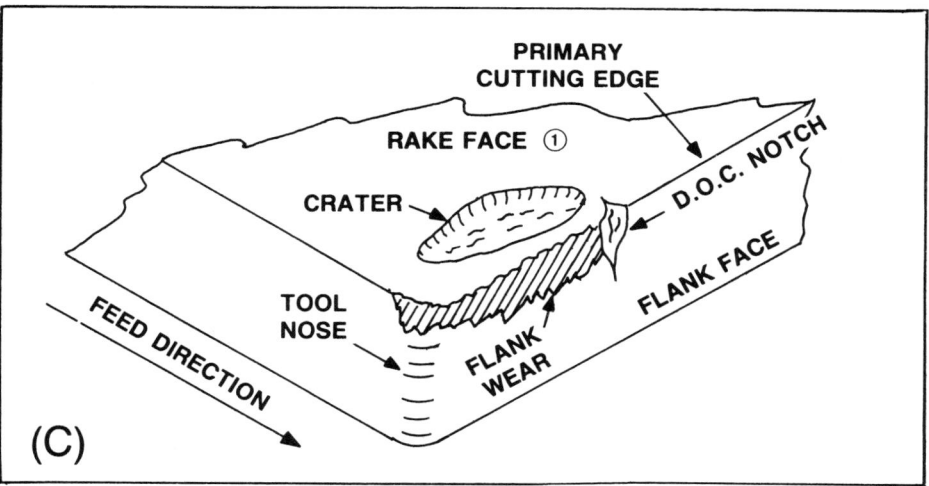

Figure 9.2 (continued) Note: D.O.C. = direction of cut.

employed. The tool bit experiences wear due to mechanical abrasion and due to chemical actions. The abrasive wear of brittle cutting tools is approximately inversely proportional to $K_{1c}^{3/4}H^{1/2}$, where K_{1c} = fracture toughness and H = hardness. In addition to the abrasive wear, chemical reactions between the tool, the work, and the ambient environment lead to dissolution and weakening of the tool bit material, contributing to overall wear. An extreme case of chemical wear occurs when silicon carbide tools are used to machine ferrous alloys. The reaction products are FeSi and FeC. The reaction between Fe and the SiC is too rapid to permit the practical use of SiC tools against ferrous alloys.

TYPES OF CERAMICS

The development of ceramics for these applications has taken four routes: (1) alumina (Al_2O_3)-based ceramics, (2) silicon nitride (Si_3N_4)-based ceramics, (3) superhard materials (diamond and Borazon), and (4) ceramic coatings. As early as 1905, alumina was used as a cutting tool. These tools were most successful in machining cast iron and high-speed finishing operations on steels at light feeds and shallow depths of cut. Alumina is limited by its relatively low thermal conductivity and its low fracture toughness.

About 1967, an improvement in alumina was achieved by the incorporation of 25 to 40 vol % of fine TiC particulates in the alumina body. Increased thermal conductivity and fracture toughness were achieved. These tools were introduced to the mass production of cast iron and rough machining of superalloy parts. Today, transformation-toughened alumina is used in which a partially stabilized tetragonal metastable ZrO_2 component is added. As described earlier, upon the application of a sufficiently high local stress, the tetragonal ZrO_2 transforms to the stable monoclinic form and thereby adds considerable toughness to the body. However, since this toughening mechanism is temperature dependent, decreasing as the temperature increases, the application of the transformation-toughened alumina is limited to lower feeds and speeds, where the tool does not get too hot. SiC whisker-toughened alumina also exhibits increased toughness and is not as temperature sensitive as the ZrO_2-alumina. This material is now most favored for rough machining of the superalloys. Because of the SiC reaction with Fe, these tools are not suitable for use on ferrous alloys.

Silicon nitride-based cutting tools are finding favor in the machining of cast iron. Since Si_3N_4 is virtually impossible to form by solid-state sintering, Si_3N_4 usually contains a variety of sintering aids. Aluminum and oxygen are added to form a material with the

CERAMIC CUTTING TOOLS

acronym SiAlON (silicon-aluminum-oxygen-nitrogen). This material contains some glassy phase content and the performance is highly dependent on the exact composition and processing route. Table 9.1 compares alumina and Si_3N_4-based tools in the machining of cast iron. For this application (and lots of cast iron is cut every day) Si_3N_4-based tools are clearly superior.

In high-speed finishing and semiroughing operations the silicon nitride-based materials have been found to outperform alumina-based tools as well as alumina-coated cemented carbide tools. An additional advantage of Si_3N_4-based tools is that they perform even in older machine tools. In general, machine tools have to be built with much more rigidity in order to inhibit fracture of the ceramic and to take full advantage of the new ceramic tools.

The superhard materials are diamond and cubic boron nitride (e.g., Borazon). Cubic BN has the same crystal structure as diamond and is the second hardest of all materials (Knoop hardness 4700 kg/mm^2), diamond being the hardest (Knoop hardness 8000 kg/mm^2). Cubic BN is about half as hard as diamond. Strictly speaking, diamond is not a ceramic. It is a form of the pure element carbon. Cubic BN, however, is a ceramic. Table 9.1 shows the hardness of various hard materials and their relative erosion resistance.

Table 9.1 Erosion Resistance Versus Hardness

Material, in increasing order of erosion resistance	Knoop hardness (kg/mm^2)
MgO	370
SiO_2	820
ZrO_2	1160
Al_2O_3	2000
Si_3N_4	2200
SiC	2700
B_4C	3500
Diamond	7000–8000

(a)

Figure 9.3 (a) Polycrystalline diamond cutting tool inserts mounted in holders. (b) Advanced polycrystalline Al_2O_3 tool inserts and holders.

(b)

Diamond has been used for cutting applications since antiquity. Despite its extreme hardness, its use as a cutting tool is limited because diamond reacts in air above 500 °C. Diamond reacts with carbide to form such metals as Fe, Ti, Zr, W, Co, and Ni by chemical reduction. At 1000 °C diamond reacts with iron. Initially, single-crystal diamond was incorporated in a steel shank to form the cutting tool. The cutting edge could be lapped to an extremely fine finish. Surfaces with less than 10-μin. roughness were thus achieved. The tool was used on nonferrous metals, where light cuts could be taken to avoid chipping and fracturing of the crystal.

An improved cutting tool was developed aroung 1972, which employed a sintered polycrystalline diamond compact, typically 0.5 mm thick bonded to a substrate. The bonding requires much the same pressure (about 50 kilobar) and temperature (about 1500 °C) as that used to synthesize diamond from graphite. The polycrystalline diamond can take higher stresses than the single-crystal diamonds. A cutting tool is fabricated by brazing the diamond-coated substrate to a toolholder. Figure 9.3(a) shows typical polycrystalline diamond cutting tools. Although these tools are far more expensive than cemented carbide tools, in some applications the polycrystalline diamond outperforms the cemented carbide in terms of tool life by a factor of 100 to 1. Such an application is the machining of a high-silicon-content aluminum alloy, where the abrasiveness due to the silicon rapidly wears other types of tools. However, the cost effectiveness of diamond occurs only for specialized applications. Cubic BN does not react with ferrous alloys. It is manufactured and used in a similar fashion as polycrystalline diamond. Cubic BN is cost-effective for steels with Rockwell hardness above 45 and superalloys with Rockwell hardness above 35. Figure 9.3(b) illustrates an advanced alumina cutting tool.

CERAMIC-COATED TOOLS

Ceramic-coated tools are also cost-effective for many applications. The first ceramic-coated tool was a cemented multicarbide tool coated with a thin TiC layer which acted as a diffusion barrier between the tool and the workpiece, thus retarding wear due to chemical reactions.

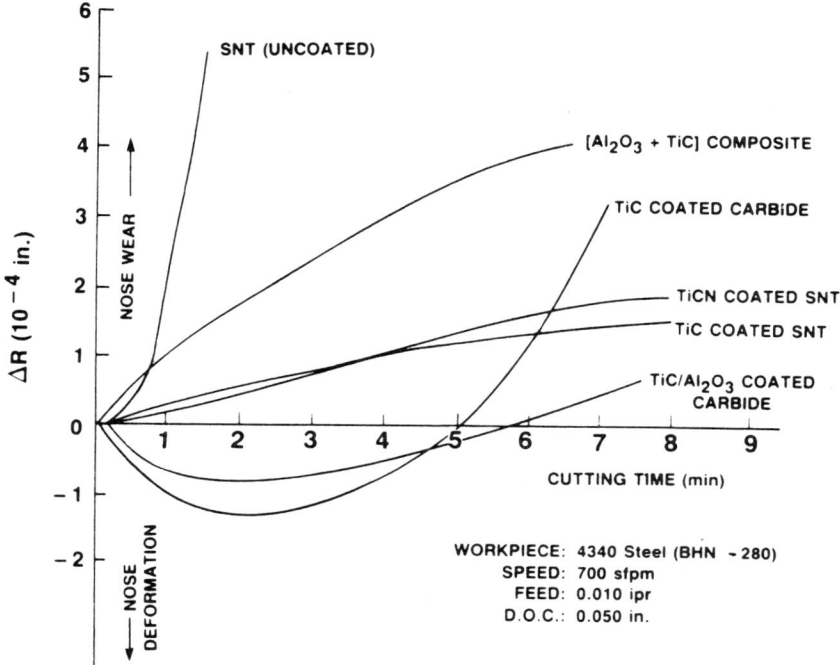

Figure 9.4 Tool wear ($+\Delta R$) and deformation ($-\Delta R$) of a selected series of cutting tools (SNT = Si_3N_4 + 6 wt % Y_2O_3 + 2 wt % Al_2O_3 + 30 vol % TiC).

Table 9.2 Comparison of Metal Removal Rates of Various Cutting Tool Materials for Gray Cast Iron Machining

Cutting tool material	Speed		Feed (per revolution)		Depth of cut		Metal removal rate			
	m/s	sfpm	cm	in.	cm	in.	m^3/s	in.^{3}/min	kg/s	lb/min
Al_2O_3-coated cemented carbide[a]	6	1200	0.025	0.010	0.318	0.125	0.00005	18	0.03	4.6
Al_2O_3[a]	15	3000	0.013	0.005	0.152	0.060	0.00003	10.8	0.02	2.8
Al_2O_3 + TiC[a]	15	3000	0.025	0.010	0.152	0.060	0.00006	21.6	0.04	5.5
Si_3N_4-based[b]	25	5000	0.076	0.030	0.478	0.188	0.00091	337.5	0.65	86.3

[a]Catalog-recommended conditions.
[b]Field data.

A wide variety of commercially available coatings exist including:

Single layer	Laminates
TiC	TiC-TiN
TiN	TiC-Al$_2$O$_3$
HfC	TiC-Al$_2$O$_3$-TiN
HfN	

In the laminates, the TiC is used as a first layer to promote bonding to the carbide tool. Coated tools comprise about a third of the cemented carbide tool market. Figure 9.4 shows the relative performance of various tools in machining 4340 steel (DOC = depth of cut). Table 9.2 shows the general trends of the properties and performance of the various types of cutting media.

10
Space Shuttle Insulation Tiles

A dramatic and epoch-making development of lightweight reusable insulation materials was accomplished during the design and construction of the Space Shuttle orbiter by NASA. This development took six years of intensive effort and was a necessary precursor to successful Shuttle operation. Figure 10.1 shows the orbiter and the temperatures reached during reentry in a typical trajectory. During reentry of the Shuttle into the earth's atmosphere, its surface reaches 2300 °F where the ceramic tiles are used. Even hotter regions occur at the nose tip and airfoil leading edges, where graphite composites must be employed. Whereas the carbon composites ablate during reentry, the ceramic tiles merely get hot, with no material loss at all. Barring mechanical damage to the delicate tiles, they can survive 100 flights without replacement due to degradation of the ceramic fibers and binders from which the tiles are composed.

INSULATION TYPES FOR THE SHUTTLE ORBITER

The basic material used for shuttle insulation is a very fine glassy fiber of silica (SiO_2). Figure 10.2 shows the approximate layout of

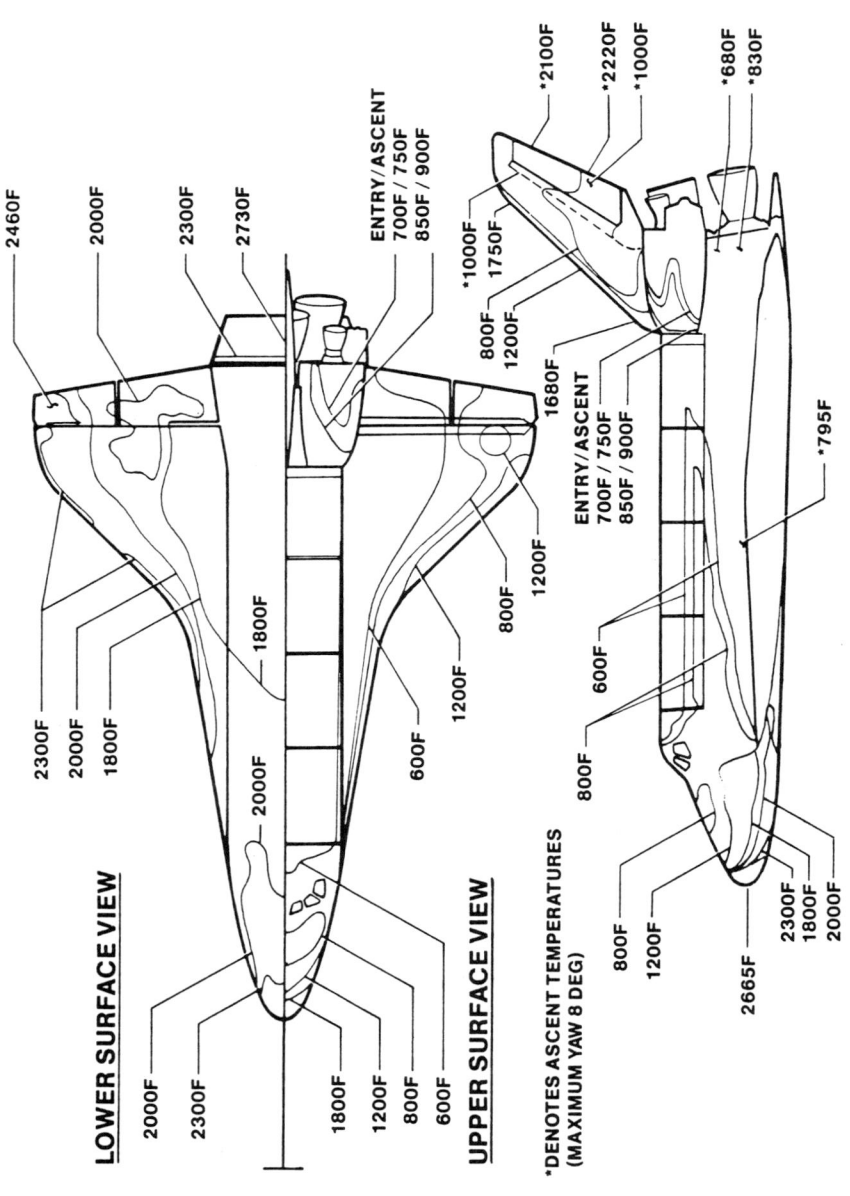

Figure 10.1 Orbiter Isotherms-Trajectory 1441.1C.

SPACE SHUTTLE INSULATION TILES

- **TOTAL RSI CERAMIC TILES - 30,812 (LMSC 24,500)**
- **REINFORCED CARBON/CARBON (RCC) (44 PANELS/NOSE CAP)**
- **FELT REUSABLE SURFACE INSULATION (FRSI) (3,581 FT^2)**

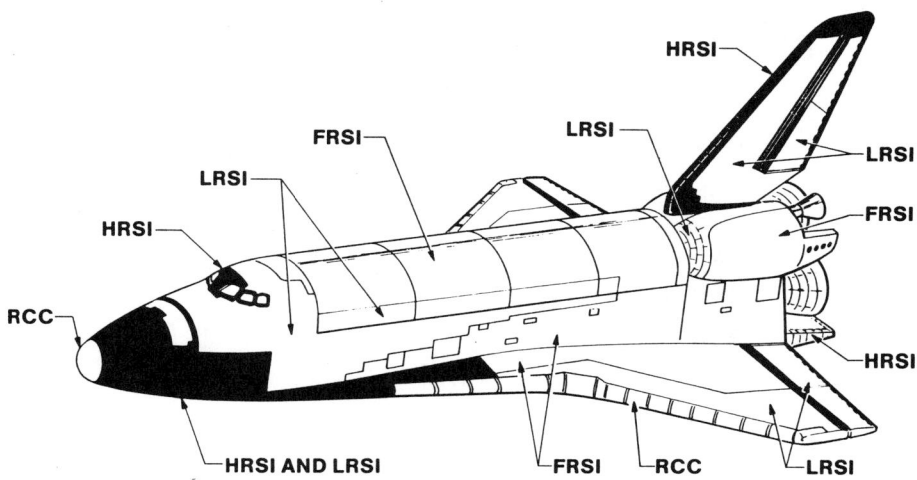

Figure 10.2 Columbia (OV-102) TPS (Thermal Protection System) Locations.

the tiles on the orbiter surface. HRSI is the high-temperature tile designed for the 2300 °F temperature regime. LRSI is the low-temperature tile designed for temperatures in the range 750 to 1200 °F. The insulation protects the aluminum structure of the vehicle so that the aluminum never exceeds 350 °F. The HRSI and LRSI differ only in the surface coating. The high-temperature material is coated with 15 mils of a high-emissivity black reaction-cured borosilicate glass. The high emissivity serves to reradiate the heat efficiently during reentry. The low-temperature material is coated with a white silica/aluminum oxide coating designed to reflect the sun's radiation during the orbiting phase of the mission. The areas of the surface that are heated less severely get the white coating.

A relatively small portion of the orbiter is covered with a stronger material. Where wear and tear may be a problem, such as access doors, main landing gear door, and other limited areas, a more rugged material is needed. One type is the LI-2200, which is about 22 lb/ft^3 in density. A newer type is designated FRCI-12 and is only 12 lb/ft^3 in density. This material is made from a mixture of 80%

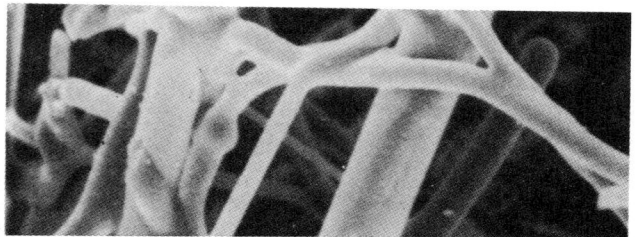

FRCI @ 2420°F

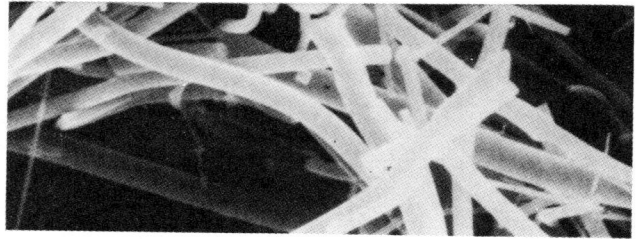

LI @ 2350°F

Figure 10.3 Fiber fusion characteristics.

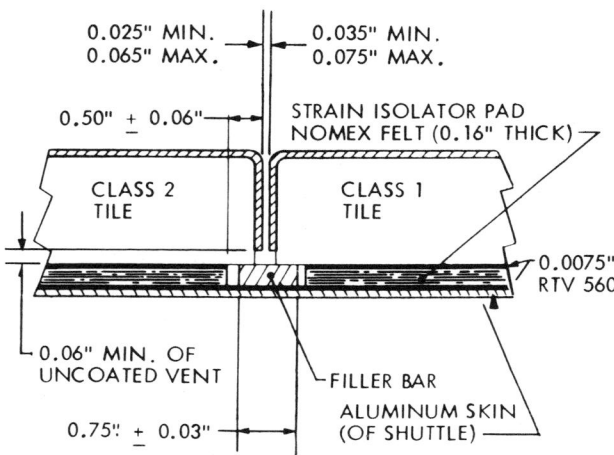

Figure 10.4 Insulation placement.

SPACE SHUTTLE INSULATION TILES

Table 10.1 Typical Physical Properties

	LI-900	LI-2200	FRCI-12
Density (lb/ft^3)	8.0-9.5	20-24	11.9-13.5
Tensile strength* (lb/in^2)			
Thru-the-thickness	24	73	81
In-plane	67	180	257
Compressive strength* (lb/in^2)			
Thru-the-thickness	40	130	160
In-plane	95	230	185
Thermal expansion* (in/in-°F)			
Thru-the-thickness	4×10^{-7}	4×10^{-7}	7×10^{-7}
In-plane	4×10^{-7}	4×10^{-7}	7×10^{-7}
Apparent thermal conductivity* (BTU-in/ft^{2}hr-°F)			
Thru-the-thickness			
70°F @ 10^{-4} ATM	0.10	0.32	0.13
1000°F @ 10^{-4} ATM	0.31	0.44	0.34
In-plane			
70°F @ 1 ATM	0.37	0.68	0.53
1000°F @ 1 ATM	0.79	1.22	1.13
Specific heat* (BTU/lb-°F)	0.17	0.17	0.17

*Average value

pure silica fibers and 20% Nextel containing a small amount of boron, which causes the silica and Nextel fibers to fuse during the thermal treatments, strongly bonding the fibers to each other and giving the mass about the same robustness as the more loosely bonded LI-2200. Figure 10.3 shows the two types of microstructure. Note the "welds" between the fibers in the FRCI.

ASSEMBLY AND PROPERTIES

Figure 10.4 shows the assembly of the tiles on the skin. RTV 560 is a silicone adhesive. The strain isolator, as its name implies, isolates the strains of the aluminum skin from the relatively rigid tile to avoid cracking the tile during deflections of the orbiter structure. The tiles are small, on the order of 6 in.^{2}, because larger tiles would be cracked during normal deflections of the airframe. Table 10.1 lists properties of the various insulation materials used on the orbiter.

11
Superconductive Ceramics

One of the most surprising developments in the field of ceramics during the 1980s has been the emergence of ceramics as the best of superconductor materials. High-temperature superconductors are expected to spawn new families of applications. In this context, high temperature means temperatures above the boiling point of liquid nitrogen, 77.4 K. Thus, in 1987, when the perovskite ceramic superconductor barium-yttrium-copper-oxide ($YBa_2 \cdot Cu_3 \cdot O_7$) was discovered by P. C. W. Chu at the University of Houston to have a superconductivity onset at 93 K, the announcement immediately was recognized as bringing in a new era in superconductivity applications. Room-temperature superconductivity, if achieved, would be a breakthrough on a par with the invention of the transistor. Superconductivity in ceramics was first discovered by Bednorz and Muller in 1985, who were awarded Nobel prizes for this work. They discovered that the compound La_2CuO_4 exhibited superconductivity at 35 K, about 12 K higher than the previously highest temperature superconductor known.

In prior superconductivity work, the metals had the highest transition temperatures. No metal had been found with a transition higher than 23 K (Nb, Ge). Prior superconductors had to be cooled by liquid helium (boiling point, 4.5 K). Cooling by means of liquid nitrogen is about one-tenth the cost of cooling by liquid helium.

SUPERCONDUCTIVITY

Superconductivity is defined as the ability of a material to transmit electrical current with zero resistance and zero power loss. Thus if a coil is superconducting and a current is induced in the coil, the current would circulate forever with no energy addition after the source is removed. Figure 11.1 plots electrical resistance against temperature. A nonsuperconductor has some resistance at all temperatures. However, the resistance of a superconductor falls to zero abruptly at the critical temperature, T_c. However, in the new ceramic superconductors the drop is not abrupt. It is more typically as shown in Figure 11.1. Hence T_c has been defined as the temperature where the extrapolated ordinary curve intersects the vertical line drawn through the midpoint resistance after the discontinuity.

At T_c other properties change abruptly, for example the specific heat, as indicated in Figure 11.2. The magnetic properties also change. Below T_c magnetic fields are deflected around supercon-

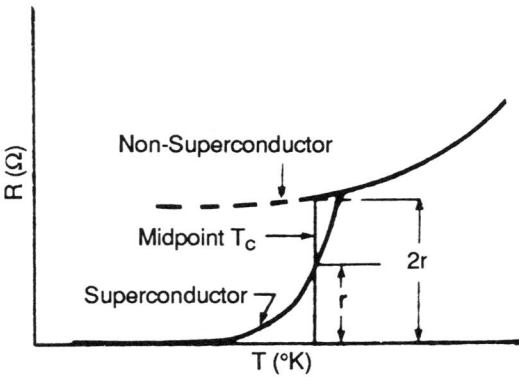

Figure 11.1 Resistance versus temperature.

SUPERCONDUCTIVE CERAMICS

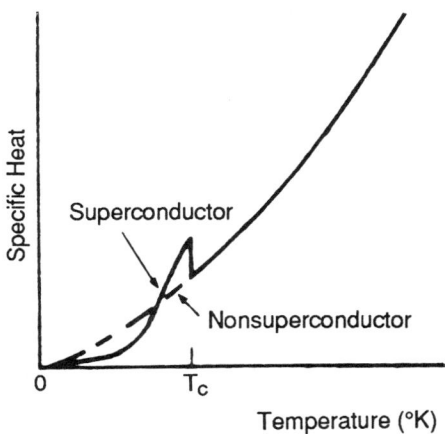

Figure 11.2 Specific heat versus temperature.

Figure 11.3 Magnetic field lines are deflected by a superconductor.

ducting substances, as indicated in Figure 11.3. A consequence of this is that a magnet will feel a repelling force if the magnet is placed near the superconductor. By appropriate selection of sizes a magnet will be levitated by the superconductor.

Another important parameter in evaluating the usefulness of a superconductor is the critical current density, that is, the maximum current density that the superconductor can carry without losing its superconducting characteristic. The $YBa_2CU_3O_7$ materials have been demonstrated to be able to carry up to 1100 A/cm², equivalent to copper at room temperature. However, this value is believed not to be an intrinsic property but rather a function of the exact formulation, processing, and impurity levels in the materials. In fact, in single crystals, current densities up to 10^6 A/cm² were achieved.

A serious effort is under way to make wire out of the superconducting materials. One approach is to place the superconducting ceramic into a silver tube and then cold drawing the tube to a fine wire. However, these wires have a low critical current-carrying capability.

JOSEPHSON JUNCTION

Another observed effect is the Josephson junction effect. Here, two superconductors are weakly coupled by a thin oxide coating. The oxide film, though insulating, has relatively low resistivity. Current flows between the two superconducting materials by tunneling. *Tunneling* is the term used for electron flow through a potential barrier that is not predictable by classical physics but is statistically possible in the quantum mechanical concept. The current flow can be affected by the application of magnetic fields. Thus when electromagnetic radiation passes through the junction, currents can be detected. These effects can lead to a variety of devices, including extremely sensitive electromagnetic detection.

APPLICATIONS

Applications for the new high-temperature superconductors are numerous and in some cases may be revolutionary. Among these applications are the following:

- Electric power storage by the use of large superconducting coils
- Extremely powerful magnets with no energy consumption
- Power generation through the use of superconducting magnets, generators, and electromotors
- Resistance free electrical transmission lines
- Magnetically levitated trains
- Electric autos powered by batteries and superconducting motors
- Specialized electromagnetic detectors
- Microelectronics and computers using Josephson junctions and zero-resistivity electrical interconnects

SUPERCONDUCTIVE CERAMICS

FABRICATION METHODS

The new superconducting ceramics are relatively easy to fabricate. Conventional ceramic processing is employed. Many of the compositions have been published, others are being covered by patents, and still others are being held as trade secrets. There is an international race on to exploit this new technology. Superconducting theory has not kept pace with practice. The theoretical basis for superconductivity is not well understood, although many scientists are trying to model the physics of superconductivity.

For the so-called 123 (one-two-three; one Y-two Ba-three Cu) material ($YBa_2Cu_3O_7$), the preferred route is by production of pure starting materials via chemical precipitation methods. The starting powders are Y_2O_3, $BaCO_3$, and CuO. A calcining process at 900 °C after milling is required. The powder is then cold pressed and sintered at a temperature in the range 950 to 1100 °C. Flowing air or O_2 is used to conserve the stoichiometry. The role of oxygen is critical to the formation of a superconductive crystal structure. An appropriate cooling rate is also required to preserve the superconducting structure. The processing is quite sensitive and each group working in this area has developed specific and detailed processing routes.

The $YBa_2Cu_3O_{7-x}$ has a perovskite crystalline structure with the basal plane consisting of Cu and O atoms and is layered so that Ba and Y atoms are stacked in the c-axis direction of the unit cell. Figure 11.4 shows the structure as revealed by an ultrahigh-powered imaging device. The perovskite structure is shown in Figure 11.5 for the perovskite crystal ($CaTiO_3$). Researchers are modifying the structure to learn further what factors in the structure are critical to the superconducting effect. There appears to be some evidence that a less than perfect perovskite structure is needed for optimum superconducting properties. The exact nature of these imperfections is being determined. The better way to describe the superconducting species is $YBa_2Cu_3O_{7-x}$ where x may be between 0.1 and 0.5. As currently understood, the superconductor structure of the $YBa_2Cu_3O_{7-x}$ is depicted in Figure 11.6. Almost any rare earth may substitute for the Y and the crystal will still exhibit superconducting properties. It is believed by some researchers that the exact nature of the bonding

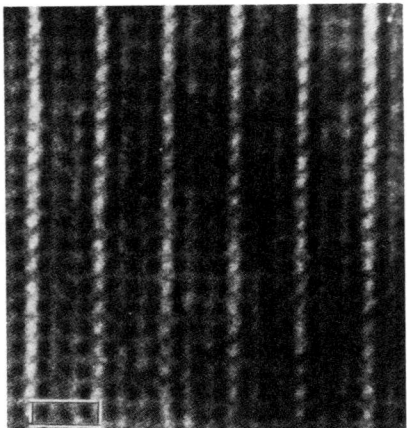

Figure 11.4 Atomic structure of high-temperature superconductor. Shown here magnified ≈ 20 million times is the first picture, taken by IBM scientists, of the atomic structure of the new high-temperature superconductors. The vertical columns of ligher spots consist of copper and oxygen atoms; they are flanked by dark vertical columns of barium atoms. Inside the columns of barium atoms are atoms of yttrium, a rare earth element. The box encloses three atoms that form the basic unit cell, which is repeated throughout the material. The box is 1.2 nm long.

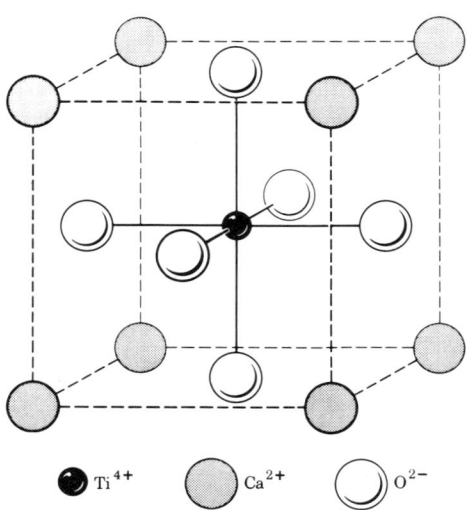

Figure 11.5 Perovskite structure (idealized).

SUPERCONDUCTIVE CERAMICS

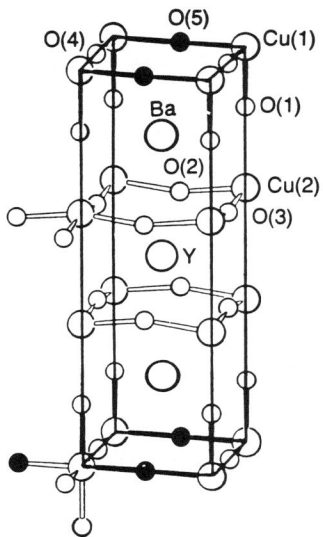

Figure 11.6 Crystal structure of $YBa_2Cu_3O_x$ showing the Cu-O-Cu-O chains, CuO_2 planes, and the ordered oxygen vacancies in the CuO_2 plane.

between the O and the Cu is a highly sensitive parameter. Note in Figure 11.6 the blackened circles where the oxygen ions should be. The blackened circles in the diagram represent vacancies, that is, vacant sites where oxygen is positioned in a normal perovskite.

During processing, conditions are maintained to ensure maximum saturation of the orthorhombic lattice with oxygen ions ($O_{7-x} = O_{6.9}$), which causes maximum distortion of the orthorhombic lattice and leads to high-temperature superconductivity, as discovered empirically. If oxygen content is increased any further, the structure transforms to another form and the superconductivity is lost. These are subtle changes and thus the processing conditions must be tightly controlled to achieve the desired superconductivity effect.

The $YBa_2Cu_3O_{7-x}$ is sensitive to moisture and must be protected from contact with moist atmosphere or other forms of water. The materials are also sensitive to ion or neutron beams. The energetic particles displace the oxygen ions and degrade the superconductivity effect.

A variety of processes are in development to produce superconducting ceramics by means other than conventional sintering. Among these methods are molecular-beam epitaxy, laser evaporation, magnetron sputtering, and from organic precursors. Some of the thin-film superconductors have very high current density capability, up to 10^6 A/cm^2 at 81 K.

12
Electronic Ceramics

It is fair to say that without ceramic technology the microelectronic revolution could not have happened. Almost every microelectronic circuit has a host of ceramic components integrally assembled together with the miraculous chips that process the electronic signals. Figure 12.1 shows an assembled integrated circuit board in the process of bonding a subassembly to the main board. The continued development of these ceramics has been an enabling technology to the miniaturization and increase of circuit density of integrated circuits. Figure 12.2 reveals the evolution of density and performance as a function of year at International Business Machine Corp. These ceramic components include conductors, resistors, capacitors, dielectrics, ferrites, ferroelectrics, and the substrates on which these conglomerates are mounted. Some of these components have been discussed in Chapter 3. Thick films as used in such circuitry refer to components that are micrometers thick, as opposed to thin films, which are angstroms thick.

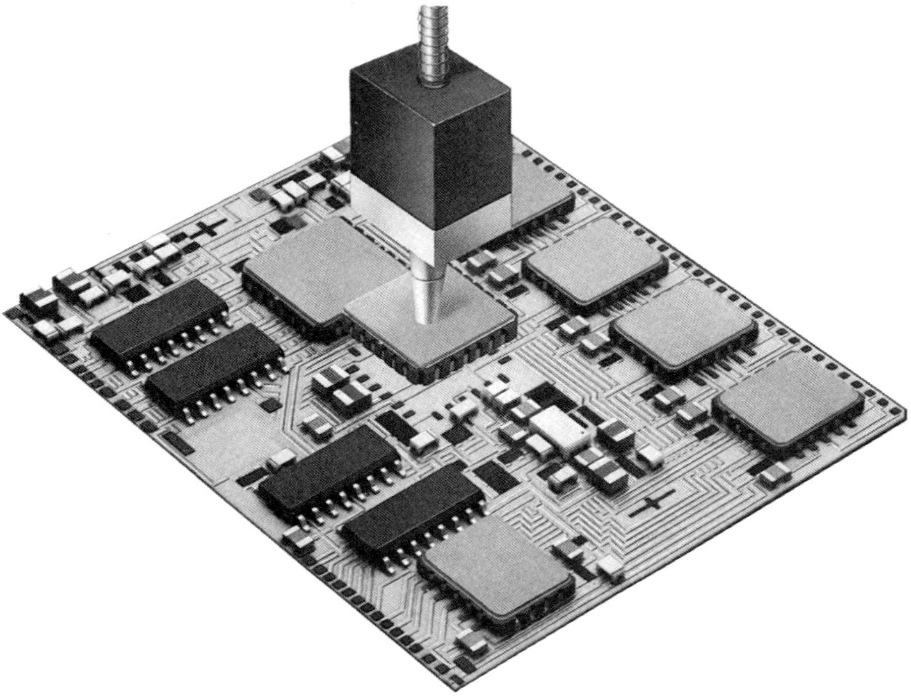

Figure 12.1

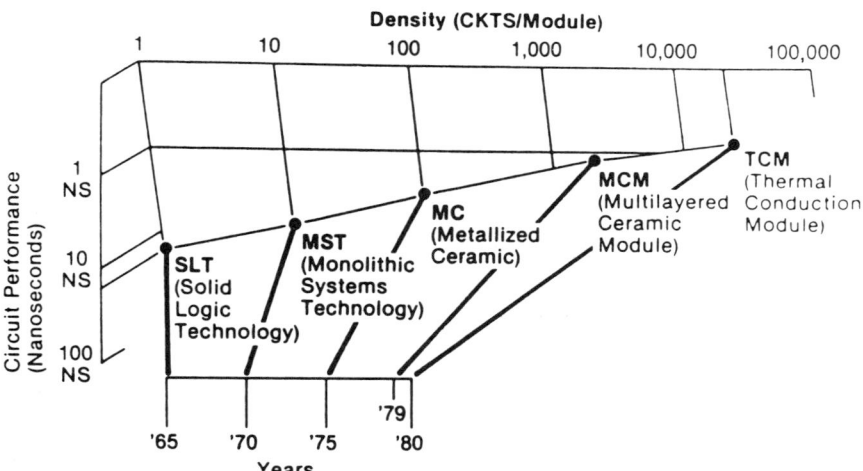

Figure 12.2 Evolution of density and performance in IBM module technologies.

CONDUCTORS

The conductor function is accomplished by the use of dispersions of fine particle metals and inorganic binders in an organic vehicle. Such inks are applied to the substrate by a silk-screen technique, dried and pyrolyzed to eliminate the organic binder. This is followed by higher-temperature firing, where the inorganic binders fuse and bond the substrate and metallic conductors. This must occur without losing the definition required. The resultant film must be adherent, compatible with the solders used to bond adjacent components, and of course have the appropriate electrical conductivity.

The very large scale integrated (VLSI) circuit fabrication places additional demands on the thick-film conductors since these have to be on the order of 1 μm wide. The conductor is typically silver and the ceramic binder is a lead borosilicate glass. The substrate is typically alumina. There are many variations on these basic ingredients. Glass content in the fired product is on the order of 10%.

RESISTOR

Resistor pastes are also dispersions of metals in inorganic binders with an organic vehicle. The ratio of metal to glass defines the resistivity. Typically, resistors contain 60 to 98% glass. Typical paste is made of a palladium silver powder in a lead borosilicate inorganic binder. The resistivity can be controlled by the proportion of metal in the binder. Fabrication is similar to the conductor process. But the resistor is usually made oversized and trimmed down by means of laser or abrasive techniques. More modern technology uses ruthenium oxide in place of the Pd-Ag mixtures. The VLSI technology is pushing the size of resistors to less than 40 mil × 40 mil with a performance equivalent to larger resistors.

DIELECTRICS

Dielectrics are used to allow conductor lines to be printed over each other (crossovers), encapsulants, and as capacitor dielectrics. The dielectrics are also made up in the form of pastes or inks and processed similarly to the conductors and resistors. Crossover dielectrics must be able to survive multiple firings without shorting out the conductors. Encapsulants should be hermetic and nondamaging to the underlying component during the processing cycle. Ideally, capacitor dielectrics should have a dielectric constant $K = 1200$ and be chemically stable.

Crossovers are either glasses with very high viscosity or compositions that contain a large crystalline content either by additions or crystallization during the processing. The high crystalline content restrains flow and possible contact between conductors during multiple firings. Firing temperatures are on the order of 800 °C. Overglaze dielectrics are generally single-phase lead borosilicate glasses with low silica and high lead content to encourage flow at low firing temperatures so as not to disturb the component being encapsulated. Such films are prone to microcracking, which remains an area needing improvement.

Capacitor dielectrics are formed by the reaction of liquid PbO and Bi_2O_3 with $BaTiO_3$. The resulting solid solution of $Pb_2Bi_4Ti_5O_{18}$ and the $BaTiO_3$ has a high dielectric constant and meets the needs of the capacitor function. Since firing is limited to 850 °C, properties are somewhat less than theoretically attainable with this composition.

13

Advanced Automotive Ceramics

The application of advanced structural ceramics in heat engines must proceed over difficult research, development, and proof testing prior to final production application. The total time span can take up to 20 years. Advanced applications of structural ceramics in automotive engines have been under intensive development since the early 1970s.

At the present time, initial trial quantities of ceramic components such as turbosupercharger turbine wheels are being evaluated in a limited number of production automobiles. One company is producing at a moderate rate a structural silicon nitride diesel engine component. Other companies are incorporating ceramic components in their engines. Catalytic converters use cordierite as a substrate for the Pt/Rh-based catalysts for pollution control. Developmental ceramic component turbine engines have been installed in U.S.-manufactured automobiles and have been track and road tested. The AGT 5 ceramic gas turbine engine was installed and road tested in a General Motors' Camaro.

Table 13.1 Characteristic Modulus of Rupture (MOR) Strength Required in Test Bars[a,b]

Probability of failure	Volume (cm³)	Weibull modulus	MOR strength required (MPa)
0.1	1.64	10	434
0.1	164.0	10	475
0.000001	1.64	10	1372
0.000001	164.0	10	2172
0.1	1.64	20	310
0.1	164.0	20	393
0.000001	1.64	20	551
0.000001	164.0	20	696
0.1	1.64	30	276
0.1	164.0	30	317
0.000001	1.64	30	407
0.000001	164.0	30	469

[a]For 207 MPa (30,000 psi) uniaxial tensile stress in the given volume for two probabilities of failure.
[b]Test bar dimensions 0.635 × 0.3175 × 3.175 cm; four-point load spans outer = 1.905 cm, inner = 0.9525 cm.

The driving force for the development of high-performance structural ceramics is the prospect for engine operation at higher temperatures with less cooling because of the refractory nature of the ceramics. The automotive market is the golden nugget at the end of the trail. The company or nation that gets there first will have a dominant position in the automotive business. The first and fundamental design issue is the assurance of structural reliability of the ceramic components.

Cost is another issue. Although ceramics are intrinsically inexpensive, when the needed reliability and performance are built into them, they become expensive. So ceramic production methods need to be highly developed for these materials to be competitive. The fuel savings are offset by the cost and failure probability of the ceramics.

For low probabilities of failure, very strong materials are needed. Table 13.1 shows data on the strength required based on stressed

ADVANCED AUTOMOTIVE CERAMICS 167

volume and probability of failure. The effect of Weibull modulus also enters into the equation. It is evident that superb materials with high Weibull moduli are required to achieve the kind of reliability we are accustomed to in our conventional metal engines.

ADVANCED GAS TURBINE PROJECT

The advanced gas turbine (AGT) 101 project, a U.S. development, aims to provide a competitive automotive engine. When fully developed and installed in a 3000-lb automobile, this engine is predicted to yield 42.8 miles per gallon, meet emission standards, and be capable of operating on alternative fuels. It is a 100-hp engine. The project has been funded by the U.S. Department of Energy and NASA and performed by the Garrett Turbine Engine Company and the Ford Motor Company.

The single-shaft regenerated engine configures all the ceramic components except one to be objects of revolution. This greatly simplified fabrication and reduces stresses in the parts. The turbine inlet temperature is designed to be 2500 °F. A cross section of the engine is shown in Figure 13.1. The ceramic components are displayed in Figure 13.2. Table 13.2 lists the ceramic construction materials and the manufacturers of the ceramics. The ceramic turbine spins at 100,000 rpm. Computed stress distribution of the turbine hub at 134,000 rpm is shown in Figure 13.3. This engine represents the most ambitious ceramic structural challenge ever undertaken. In the development of this engine, much progress was made in the design methodology and statistical design techniques. Materials have been advanced and the reduction in flaws in the ceramic parts has been brought to a new level of perfection. However, much work still needs to be done to achieve the reliability and cost targets required for the automotive markets.

One of the most severe thermal shocks the engine suffers arises during startup. Figure 13.4 shows the start transient. The ceramic combustor evolved into a multiple-piece assembly to reduce stresses. As was discussed in the Space Shuttle orbiter thermal tile design, smaller pieces of ceramic can better handle thermal shock loads. The multiple-piece design lowered maximum principal stresses to less than 20 ksi. The rotor component has focused on two kinds of silicon nitride: sintered reaction-bonded silicon nitride (SRBSN)

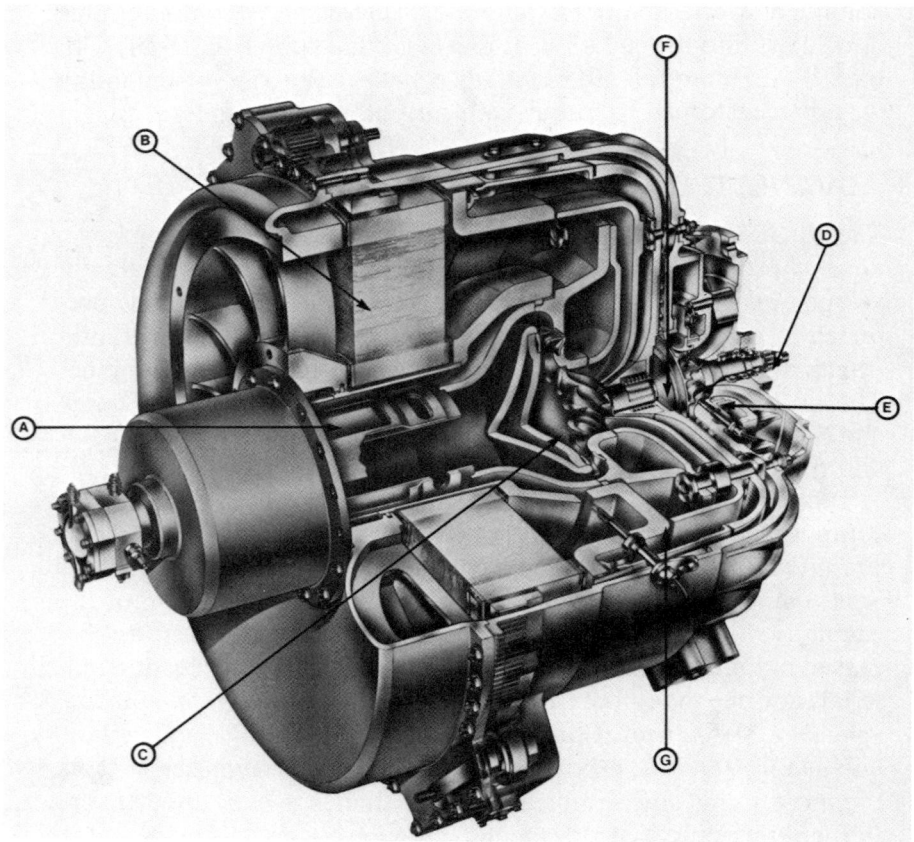

Figure 13.1 AGT 101 Power Plant. (A) combustor (B) regenerator (C) turbine (D) ball bearing (E) compressor (F) foil gas bearing and (G) ceramic structures.

and sintered silicon nitride (SSN). Figure 13.5 shows how the materials improved during the development period, with SSN showing extremely good flexure strength by 1985. These higher strengths were accompanied by an improvement in Weibull modulus, as well. As an example, a sintered silicon nitride rotor is shown in Figure 13.6.

ADVANCED AUTOMOTIVE CERAMICS

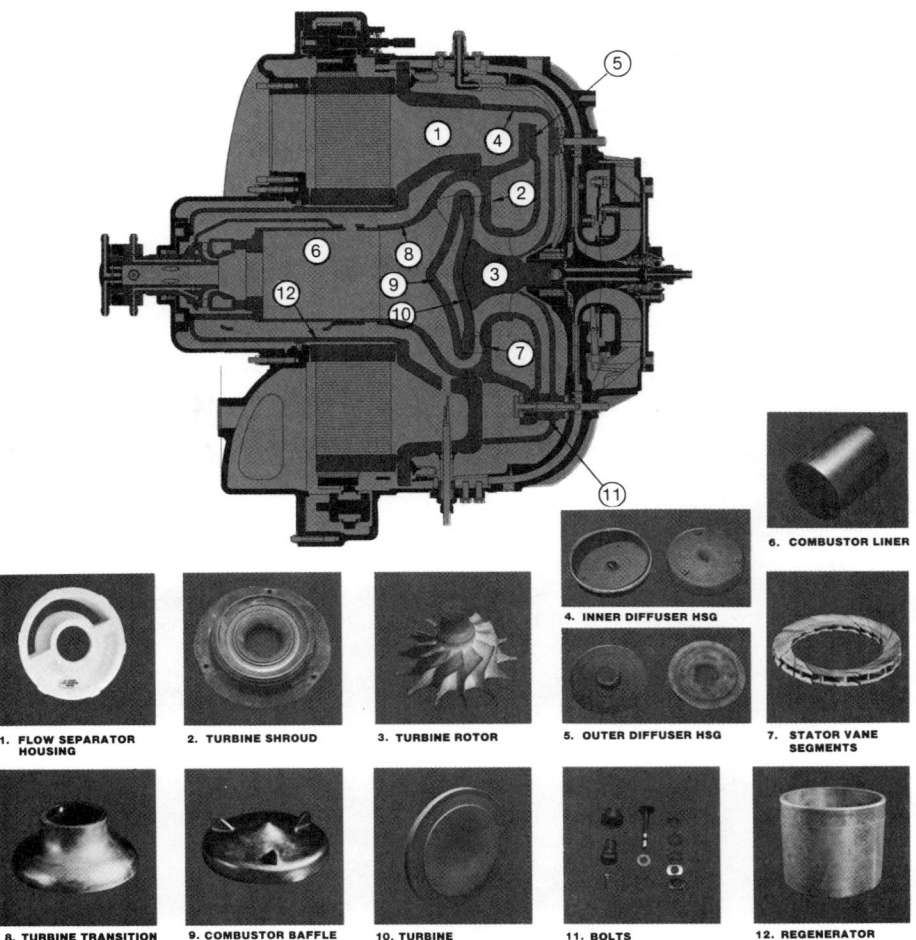

Figure 13.2 Ceramic components of the AGT 101 power plant.

Typical Development Cycle for Ceramic Component

Development of critically stressed ceramic components for the AGT 101 follows a closed-loop iterative design-fabrication-test-redesign

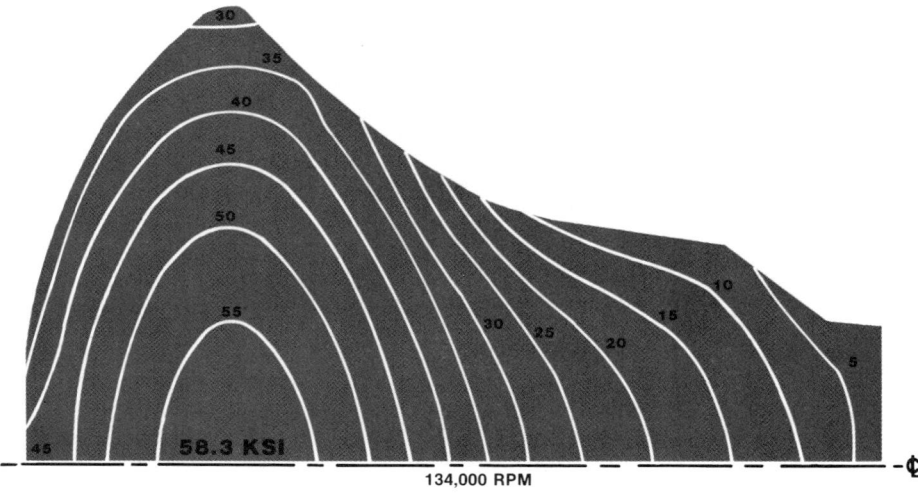

Figure 13.3 Stress distribution of Ford rotor at 134,000 rpm.

Table 13.2 Ceramic Construction Materials[a]

Part Number in Figure 13-2	Carborundum	Ford Corning	ACC	Pure Carbon	NGK
1		LAS			
2	αSiC		RBSN		
3	αSiC	SRBSN	SSN	RSSC	
4			RBSN		
5			RBSN		
6	αSiC			RSSC	
7	αSiC	RBSN	RBSN		
8	αSiC		RBSN		SSN
9	αSiC		RBSN		
10	αSiC				SSN
11	αSiC		RBSN		
12	αSiC		RBSN		

[a]SiC, sintered α-silicon carbide; LAS, lithium aluminum silicate; SRBSN, sintered reaction-bonded silicon nitride; RBSN, reaction bonded silicon nitride; SSN, sintered silicon nitride; and RSSC, reaction sintered silicon carbide.

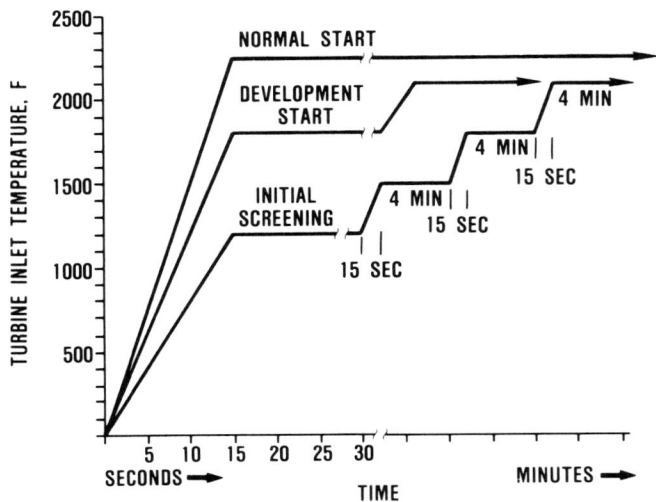

Figure 13.4 Start transient comparison.

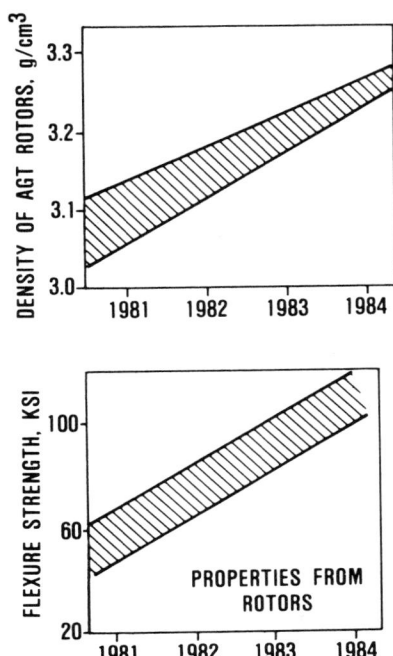

Figure 13.5 Density and strength of ceramic rotors continue to improve.

(a)

(b)

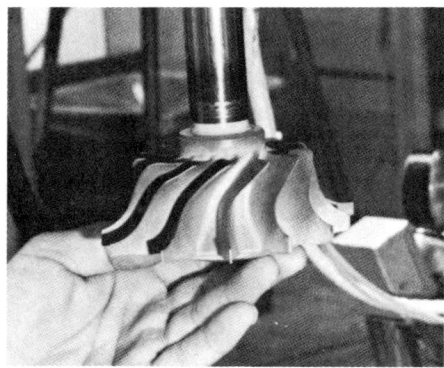

(c)

Figure 13.6 Ceramic turbine rotor from Airesearch Casting Co. (a) Ford nonbladed rotor after 134,000 RPM test, (b) bladed ceramic rotor, and (c) rotor spin tested to 102,000 RPM.

ADVANCED AUTOMOTIVE CERAMICS

approach. This procedure is represented in Figure 13.7. In this method, thermal and stress analysis are updated as more information becomes available from the results of fabrication and testing experience. At each stage of the cycle, fabrication and test are evaluated, and parts that fail are subjected to intensive failure analysis. The data are fed back to the design step, with the part being redesigned for the fabrication processes refined to create a more survivable component. When the process is completed, a part has been developed which passes the fully assembled power section development test. This iterative process typically takes years of sustained development work.

As an example, the turbine shroud design for the AGT 101 engine underwent a series of redesigns before it could withstand the thermal transient stresses developed during normal engine start. One change in the analysis was to employ a finer finite element mesh

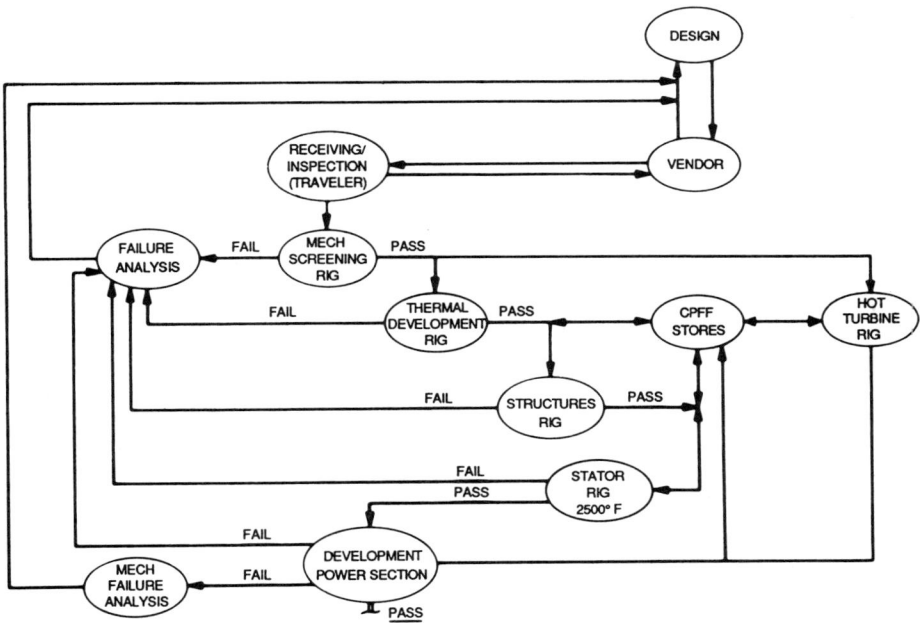

Figure 13.7 Ceramic hardware cycle.

in order to define the highly localized stresses in the corners of the structure. The shroud is shown in Figure 13.1.

Design Methodology

The original finite element stress analysis mesh is shown in Figure 13.8. The resulting stress prediction is indicated in Figure 13.9(a). When a considerably refined mesh was used, the much higher peak stress shown in Figure 13.9(b) was revealed, going from 9.9 ksi to 25.6 ksi. In the fabricated piece this stressed area was further aggravated by machining grooves in the reentrant corner. To resolve this problem, additional material was added to the seal land area, and polishing after final grind was included in the manufacturing procedure.

With these modifications, the shroud was successfully proof-tested in the screening rig to 22.4 ksi, as indicated in Figure 13.10. Another example during the development of the AGT 101 was the turbine backshroud shown in Figure 13.11, which fractured into four almost equal pieces during hot rig testing. The location of the backshroud is indicated in Figure 13.1.

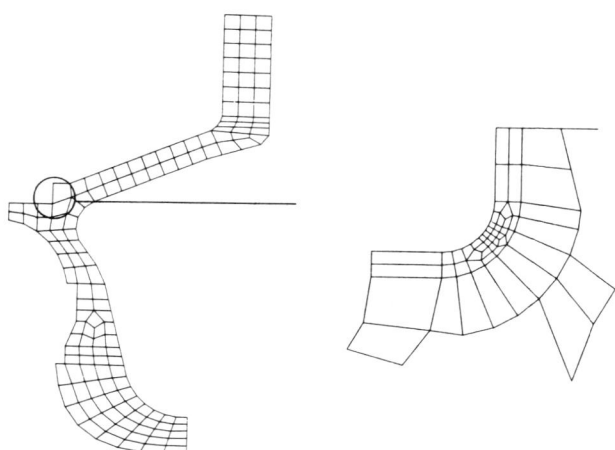

Figure 13.8 Turbine shroud/flow separator housing/transition duct seal land area as originally modeled.

ADVANCED AUTOMOTIVE CERAMICS

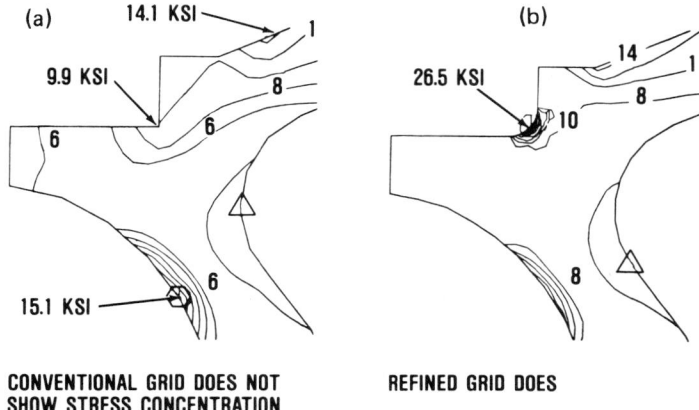

Figure 13.9 (a) Initial stress prediction, and (b) resulting stress prediction from refinement.

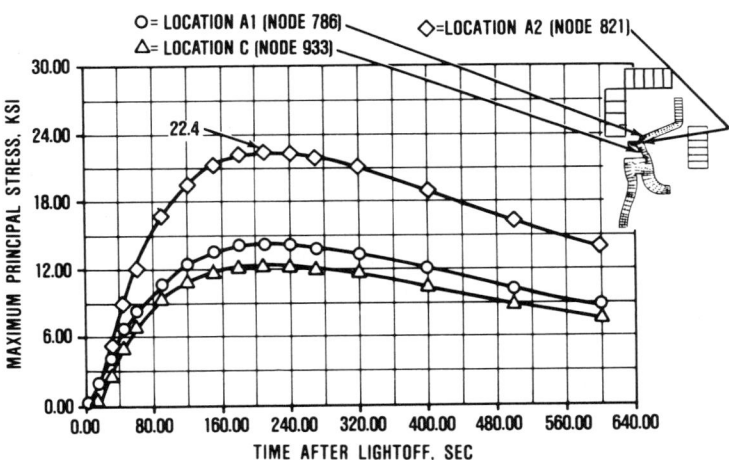

Figure 13.10 New shroud was successfully proof-tested with modifications.

Figure 13.11 Turbine backshroud fractured.

A review of the stress analysis showed that the tensile stresses could be alleviated by providing a central hole, 3 in. in diameter, in the shroud and filling the hole with the Lockheed insulating tile material described in Chapter 10. This solution worked very well, as attested by a series of five transient test cycles without a failure.

Along with refinements in the analysis, material fabrication consistency and a knowledge of the properties of the materials are essential. Progress in these areas is steadily being made. Remember, refined metallurgical processes have been in development for at least a century, but the engineering ceramics refinements required for the extreme applications now being contemplated really started in the 1950s.

A parallel gas turbine development was performed by a team of the Allison Gas Turbine Division of General Motors Corporation, its Pontiac Division, Delco-Remy, Carborundum Company, Corning Glass Works, and GTE Laboratories, Inc. The engine, designated the AGT 100, is shown in Figures 13.12 and 13.13. The regenerator is shown on the right. The opposite end is the cold sec-

ADVANCED AUTOMOTIVE CERAMICS

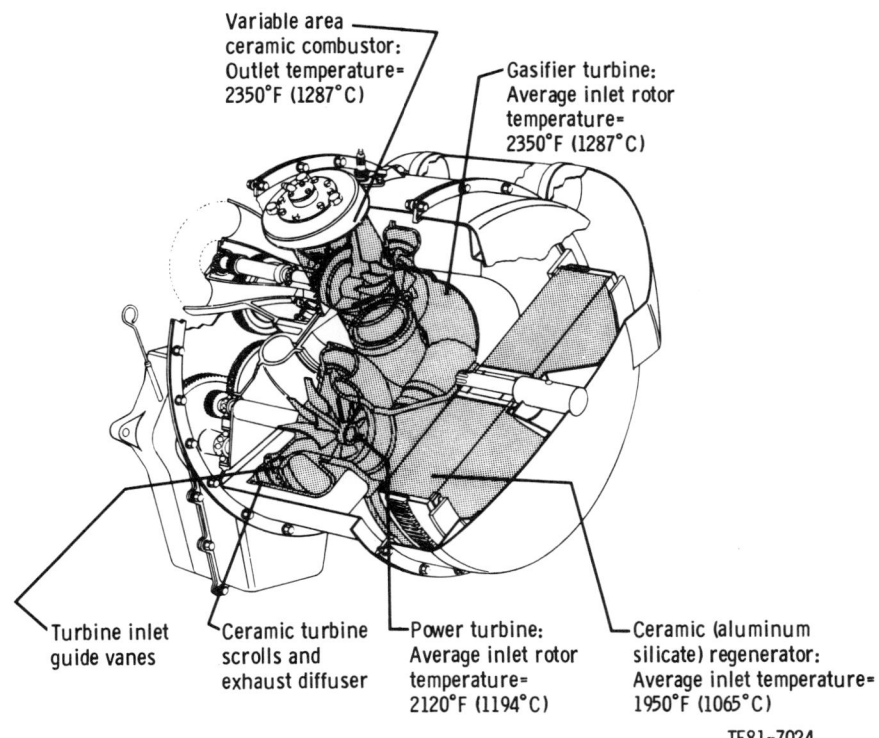

Figure 13.12 Ceramic content of the AGT 100 engine.

tion containing the air compressor and the gearbox. The center is the hot section and includes the combustor, the turbine, and the ducting, these components all being ceramic. The engine is fitted with two shafts, one for the gasifier turbine and the power turbine on the second shaft. The critical ceramic components, as shown in Figures 13.12 and 13.13, are the ceramic combustor, the gasifier turbine, the regenerator, the power turbine, the ceramic turbine scrolls and exhaust diffuser, and the turbine inlet guide vanes.

The gasifier turbine scroll is fabricated from five silicon carbide (SiC) pieces, which are then assembled using a high-temperature ceramic brazing technique. Silicon nitride gasifier scrolls have also

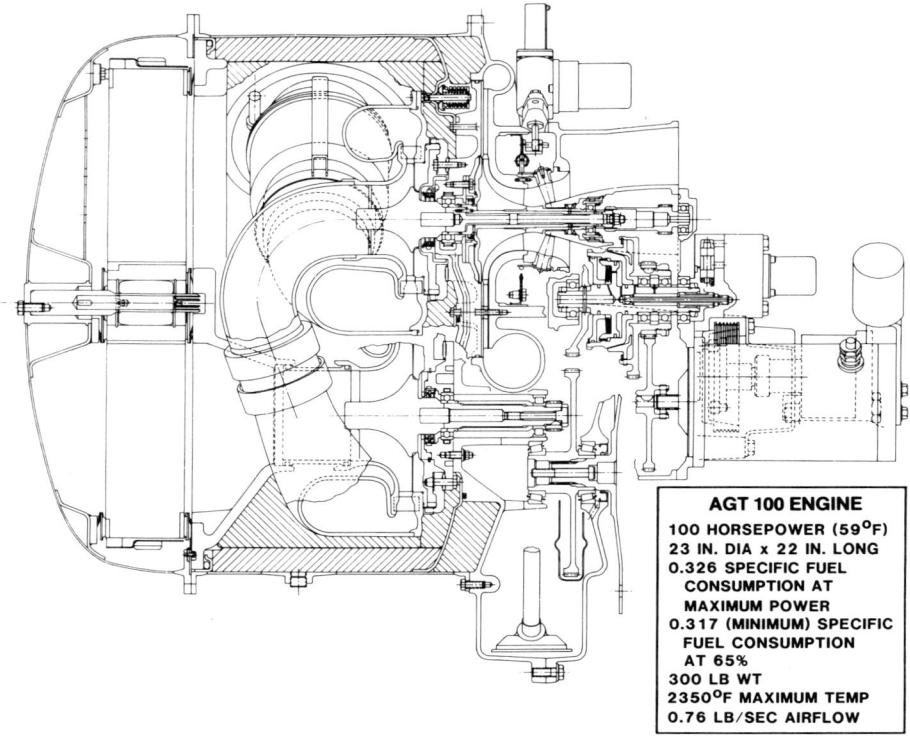

Figure 13.13 AGT 100 engine cross section.

been fabricated as part of the AGT 100 program. An assembled scroll is shown in Figure 13.14. The process for assembly of the SiC scroll is shown schematically in the flow diagram of Figure 13.15. The bond is created by the application of silicon followed by a thermal treatment. The bond is 0.001 to 0.002 in. thick. The three-dimensional finite element mesh used in the design of the scroll is shown in Figure 13.16. The three-dimensional analysis was supplemented by appropriate two-dimensional analyses which are more cost-effective to conduct. The complexity of the design-development effort is very high, requiring the very best of analytical, design, fabrication, and test and evaluation technology. Unfortunately, the

ADVANCED AUTOMOTIVE CERAMICS 179

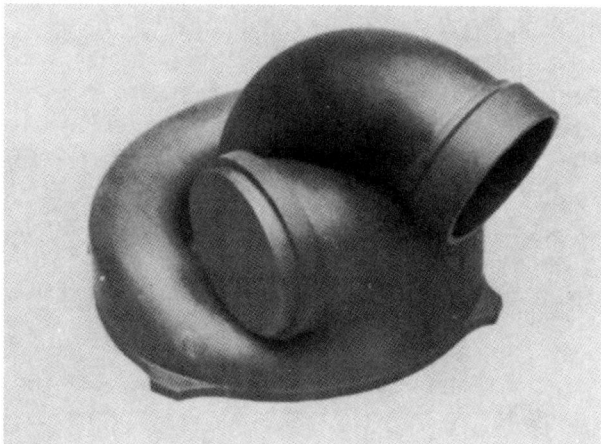

Figure 13.14 Completed scroll assembly.

finite element technique deals primarily with fast fracture. The effects of long-term usage at elevated temperatures can only be derived by extensive life testing, which simulates the varieties of service such an engine must endure.

ADIABATIC DIESEL ENGINE

The adiabatic diesel engine is an insulated engine that does not require a liquid coolant for prevention of overheating of the engine components. The materials of fabrication (i.e., ceramics) can withstand much higher temperature than can a conventional all-metallic diesel. The engine is not truly adiabatic, since that would imply no heat transfer at all between the engine and its surroundings. The design seeks maximum thermal conversion efficiency by insulation of the combustion chamber, recovery of exhaust heat by means of turbocompounding, and elimination of the liquid cooling system.

In a program carried out by Cummins Company sponsored by the U.S. Army Tank and Automotive Command, insulated engines were designed, built, and evaluated in the early to mid-1980s. The major components of the engine that were insulated were the piston crown, the cylinder liner, and the cylinder head face, that is, the portion of the head facing the combustion chamber. The methods used to achieve insulation employed compatible ceramics, with the adjacent metal being insulated. In the case of the piston, zirconia (ZrO_2) was selected because it has an expansion coefficient close to that of the cast iron piston and can withstand high temperatures, is a good thermal insulator, and is strong.

Figure 13.17 shows an assembled insulation piston. In this case the ZrO_2 is held in place by an interference fit. However, due to the sharp thermal gradient between the top of the cast iron piston

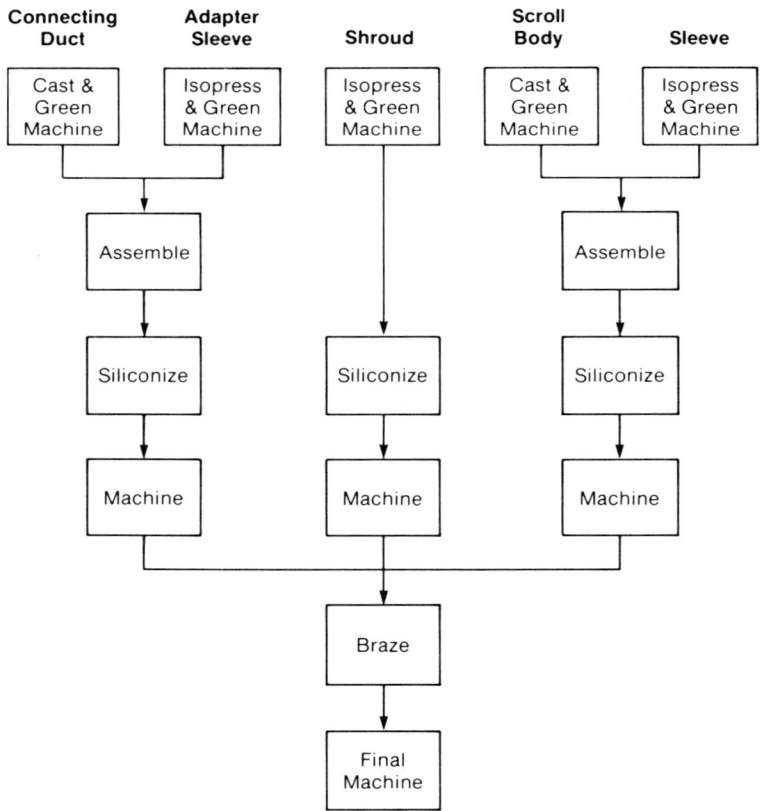

Figure 13.15 Si/SiC scroll assembly flowchart.

ADVANCED AUTOMOTIVE CERAMICS

and the surface of the ZrO_2 insulator, sufficient stress is induced in the insulator to produce cracking. Figure 13.18 indicates the types of cracks that developed during engine test. Further development is required to eliminate this type of failure.

Zirconia valve seats and valve stem guides were incorporated into the insulated engine and were very successful in testing. These components were duplicates of the metal parts and were used primarily to reduce wear. Zirconia head plates were also incorporated, shown in Figure 13.19. These head plates incorporated the valve seat and essentially were successful in engine testing. The use of zirconia as a cylinder liner was also attempted but did not operate successfully as the inserts developed cracks and excessive wear in the top piston ring. Although successive models were tested and the cylinder liner cracking was eliminated, excessive wear of the ring was observed and further development effort is required to resolve

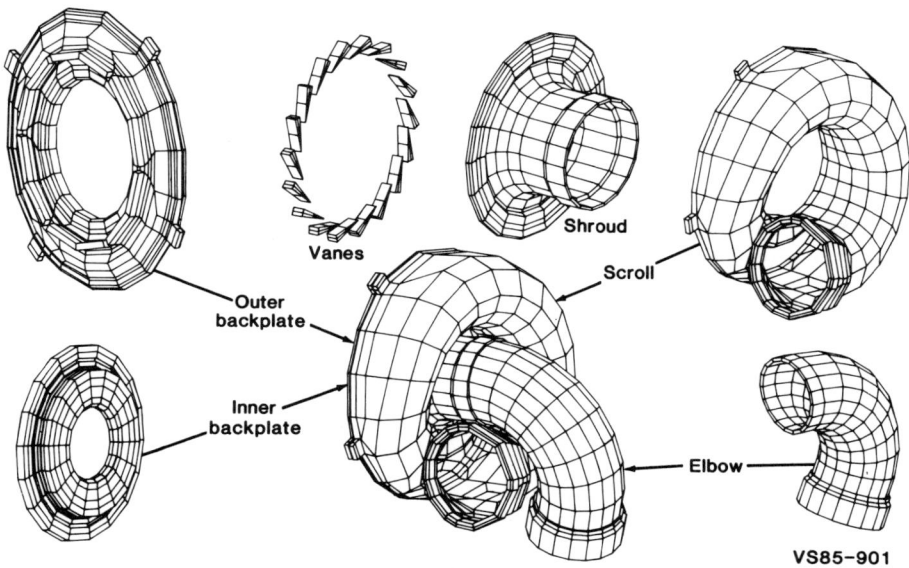

Figure 13.16 Three-dimensional finite element model RPD gasifier turbine scroll assembly.

Figure 13.17 NH engine zirconia capped ductile iron piston assembly.

that problem. The development of adiabatic engines is still ongoing using a variety of ceramic components and insulation strategies.

Current Initiatives

At the present time the Japanese automobile manufacturer Isuzu is road testing a ceramic diesel engine-powered vehicle. Cooling is by air (no water system), and 5000 km of road testing has been completed as of late 1989. The vehicle has obtained speeds of 150 km per hour. Pistons, cylinder liners, and valves as well as other components are ceramic. Again in Japan, the NGK Spark Plug Company is casting 8000 ceramic turbocharger rotors per month for the Nissan Cedric Fairlady luxury car. The main ceramic for these automotive parts is silicon nitride.

In the United States, the Corning Company makes ceramic honeycomb catalytic converter substrates for the automotive industry in large quantities. The Carborundum Co., a subsidiary of British Petroleum, is building a facility in Germany to manufacture ceramic seal rings for the European auto industry. Despite this tangible progress, ceramic science and technology has a long way to go

ADVANCED AUTOMOTIVE CERAMICS

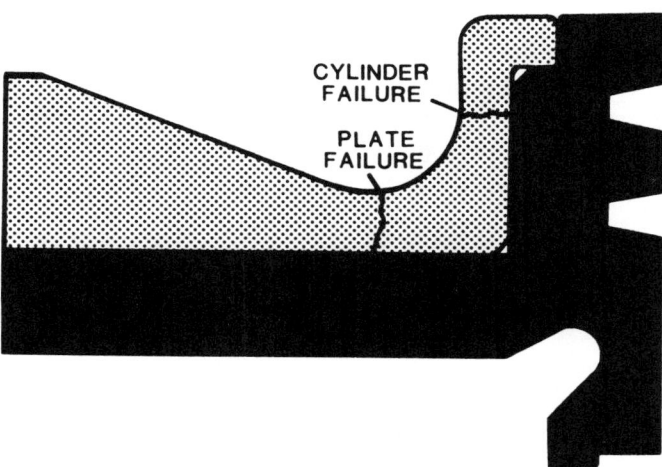

Figure 13.18 Cross-sectional view of typical NH engine piston cap failure.

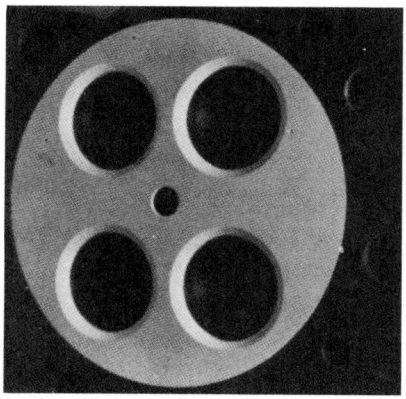

Figure 13.19 Zirconia head plate with integral seats.

before the reliability of ceramics is established, and the American company GTE Laboratories Inc. believes that the part failure rate has to be reduced from its present 1 in 10^6 to 1 in 10^9 to achieve a fully competitive state of development. A structural ceramic having a strength of 850 MPa, a Weibull modulus of 13, and a fracture toughness of 7 MPa·m$^{0.5}$ would approximate such reliability.

14
Carbon-Carbon Composites

In Chapter 10, the Space Shuttle insulation tiles were described as an example of a sophisticated application of ceramic technology. Graphite composites were mentioned as thermal protection for the nose tip and airfoil leading edges of the vehicle, required to protect the orbiter during reentry into and descent through the earth's atmosphere. During such reentry, some of the carbon oxidizes or ablates, but the amount so lost is within acceptable bounds. Such oxidization is inhibited by a ceramic coating based on silicon carbide. During reentry, the SiC reacts with oxygen and forms a silica (SiO_2) layer on the surface of the carbon. At reentry temperatures, the silica is hot and viscous and forms a diffusion barrier to oxygen, thereby slowing down oxidation of the carbon. The carbon components have to be replaced periodically, however, after a given number of flights.

Although carbon and graphite solids (like diamonds) are not, strictly speaking, ceramics it has become customery to include the advanced structural materials fabricated from carbon in the literature dealing with structural ceramics. As pointed out earlier, the incentive

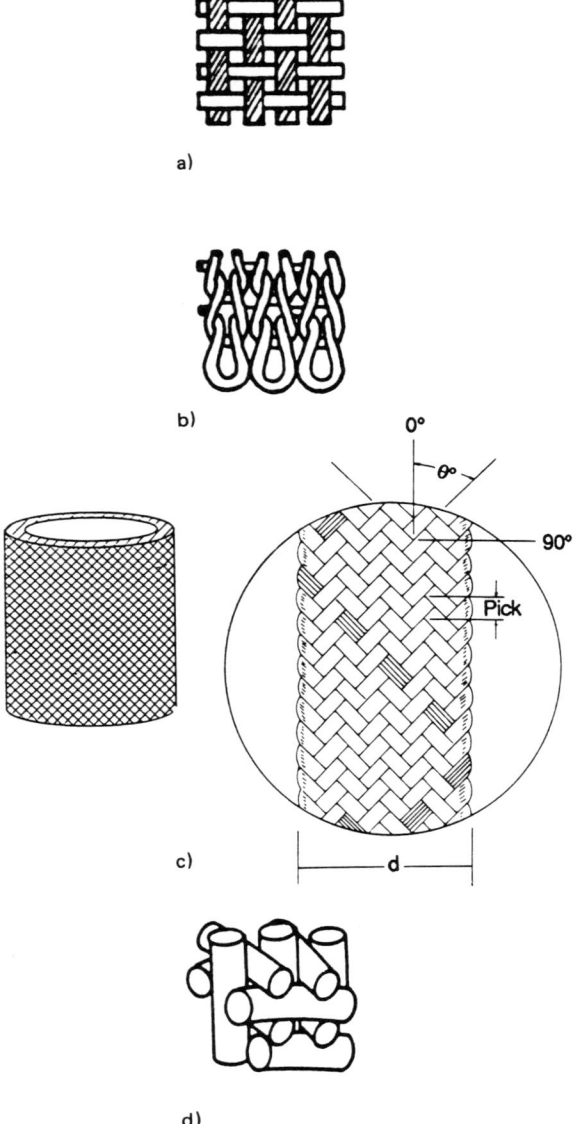

Figure 14.1 Preform architectures.

for the use of ceramics as structural materials stems mainly from the ability of ceramics to withstand higher temperatures than metals. The need for higher temperatures is driven by the Carnot efficiency in heat engines, high temperatures generated during reentry in aerospace vehicles, and high temperatures encountered in chemical reactions such as in rocket motor nozzles.

Carbon is stable at 1 atm of carbon vapor or inert gas up to 3590 °C, which is the highest temperature of stability in all the elements. In addition, the graphite crystal has a very high modulus of elasticity and a low thermal expansion coefficient (although these properties vary dramatically with orientation in the graphite crystal). The low density of graphite (specific gravity 2.2) is another advantage. All these properties provide opportunities for unique application of the materials in situations where no other material can meet the design need. The development of carbon and graphite fibers that incorporate these intrinsic characteristics of graphite and carbon and the development of the composite technology that takes advantage of these properties was an outstanding materials science and engineering achievement of the 1970s.

Aside from the all-carbon composites, a host of applications of polymers reinforced with graphite fibers also have evolved. Well-known applications are golf clubs, tennis racquets, and sailboat masts. Less well known are weaving loom components, where the combination of high stiffness and low mass helps make possible the design of extremely high speed looms. In the development of high-stiffness, high-strength composites, the advance in weaving technology has been critical to the successful tailoring of the composite properties. This is true for the polymeric as well as the all-carbon composites. The weaves used are of a great variety, being tailored to provide the needed properties in the orientations of interest. Figure 14.1 shows some of the preform architectures used.

CARBON AND GRAPHITE

Carbon bodies are conglomerates of graphite or carbon black particles which are mixed with a resin or tar, pressed, and then baked

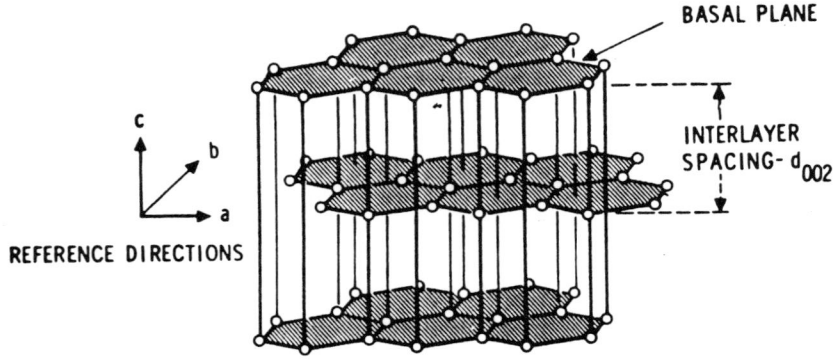

Figure 14.2 Graphite crystal.

Table 14.1 Typical Room-Temperature Properties of Some Reactor Graphites

		CS-312	AGOT	AGOT-NC	TSX
Bulk density (g/cm³)		1.71	1.71	1.72	1.73
Flexural strength (psi)	With grain	2350	2750	2100	3450
	Across grain	1850	1775	1600	1600
Tensile strength (psi)	With grain	1100	1350	1000	1400
	Across grain	1100	1050	800	800
Compressive strength (psi)	With grain	5500	5100	5100	5100
	Across grain	5700	4750	4500	4400
Young's modulus, (10^6 psi)	With grain	1.6	1.8	1.6	2.3
	Across grain	1.0	1.0	0.8	0.8
Specific resistance	With grain	8.0	6.5	5.2	5.3
(10^{-4} Ω-cm)	Across grain	12.7	10.8	10.2	12.2
Thermal expansion	With grain	1.2	1.1	0.7	0.3
coefficient (10^{-6}/°C)	Across grain	3.3	3.1	2.4	3.3

Source: Union Carbide Corp., Carbon Products Division.

CARBON—CARBON COMPOSITES

to drive off the volatile elements in the binder. Densities of such agglomerates are relatively low, and mechanical properties are relatively poor. By subjecting such bodies to temperatures up to 2500 °C (4532 °F), the graphitic structure is developed. The graphite crystal is hexagonal and tends to have low bond strength in the direction of the c-axis but is strong in the a-b (basal) plane. Figure 14.2 shows the graphite crystal. The carbon atoms in the basal plane are covalently bonded, but the layers are held together by the weak van der Waals forces. This is why the material is so highly anisotropic. Table 14.1 shows typical properties of nuclear graphites used in reactor construction.

PREFORMS AND PROCESSING

In order to develop higher mechanical properties, a class of composites generically called carbon-carbon (C-C) composites has been developed. These composites depend on a woven substrate or preform made from carbon or graphite fibers which have exceedingly high strengths and elastic moduli. Table 14.2 illustrates properties of a variety of ceramic fibers, including graphite. Fibers are available as single-filament, untwisted yarns comprised of many individual filaments, twisted yarns, and high-bulk disorganized yarn, as indicated in Figure 14.3. The preforms are made by weaving, braiding, knitting, and by nonwoven layups. These procedures are described in Table 14.3. Illustrations of preforms made by these techniques are shown in Figure 14.1.

Once the preform is prepared, it is placed in a suitable mold or retention container and the void spaces are filled with a carbonaceous tar or resin. Various thermal cycles are then performed to eliminate the volatile content and fill the voids in the preform with entirely carbonaceous materials. Multiple cycles of filling and pyrolization are required. After the desired bulk density is obtained, a graphitizing heat treatment at very high temperature on the order of 2200-3000 °C (4000 to 5400 °F) is performed to achieve the degree of graphitization desired.

Figure 14.4 shows a block diagram for this process of making carbon-carbon composites. After the carbon-carbon part is fabri-

Table 14.2 Properties of Candidate Fibers for Ceramic-Matrix Composites

Fiber	ρ (g/cm^3)	σ [MPa (ksi)]	E (GPa (Msi)]	Diameter (μm)	Maximum-use temp. (°C)
Alumina					
Fiber FP[a]	3.9	1.38 (200)	380 (55)	21	1316
PRD166[b]	4.2	2.07 (300)	380 (55)	21	1400
Sumitomo	3.9	1.45 (210)	190 (28)	17	1249
Mullite					
Nextel 440[c]	3.1	2.7 (250)	186 (30)	12	1426
Mullite/Gls					
Nextel 312[d]	2.7	1.55 (225)	150 (22)	12	1204
β-SiC					
Multifilament					
Nicalon[e]	2.55	2.62 (380)	193 (28)	10	1204
SiTiCO Tyranno[f]	2.5	2.76 (400)	193 (28)	10	1300
Si$_3$N$_4$ TNSN[g]	2.5	3.3 (362)	296 (43)	10	1204
SiC whisker VLS[h]	3.2	8.3 (1200)	580 (84)	4-7	1400
SiC monofilament					
SCS-6[i]	3.05	3.45 (500)	410 (60)	140	1299
Sigma[j]	3.4	3.45 (500)	410 (60)	100	1259
Pure fused silica					
Astroquartz[k]	2.2	3.45 (500)	69 (10)	9	993
Graphite					
T300R[l]	1.8	2.76 (400)	276 (40)	10	>1648
T40R[m]	1.8	3.45 (500)	276 (40)	10	>1648

[a] E. I. du Pont de Nemours and Co., Wilmington, DE.
[b] E. I. du Pont de Nemours and Co.
[c] 3M Co., St. Paul, MN.
[d] 3M Co.
[e] Nippon Carbon Co., Tokyo, Japan.
[f] UBE Industries, Ltd., Tokyo, Japan.
[g] Toa Nenvvo Kogyo K. K., Tokyo, Japan.
[h] Los Alamos National Laboratory, Los Alamos, New Mexico.
[i] Avco Specialty Materials/Textron, Lowell, MA.
[j] Berghof, Tübingen, Federal Republic of Germany.
[k] J. B. Stevens, Greenville, SC.
[l] Amoco Performance Products, Ridgefield, CT.
[m] Amoco Performance Products.

CARBON-CARBON COMPOSITES

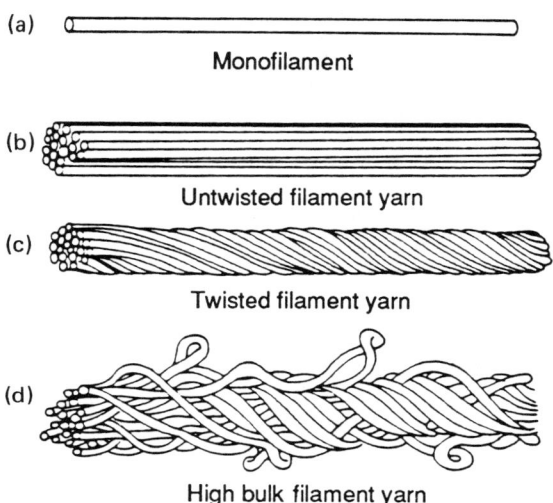

Figure 14.3 Structural geometry of filament yarns: (a) monofilament; (b) untwisted filament yarn; (c) twisted filament yarn; (d) high-bulk filament yarn.

Table 14.3 Comparison of Yarn-to-Fabric Formation Techniques

	Basic direction of yarn introduction	Basic formation technique
Weaving	Two (0°/90°) (warp and fill)	Interlacing (by selective insertion of 90° yarns into 0° yarn system)
Braiding	One (machine direction)	Intertwining (position displacement)
Knitting	One (0° or 90°) (warp or fill)	Interlooping (by drawing loops of yarns over previous loops)
Nonwoven	Three or more (orthogonal)	Mutual fiber placement

cated, it is coated with oxidation-inhibiting SiC, followed by TEOS (tetraethylorthosilicate) impregnation to seal the cracks in the SiC which form during cool-down from processing temperatures. An alternative technique is to infiltrate the preform with carbon by chemical vapor deposition (CVD) in an isothermal process. In this process, the preform is placed in a vacuum chamber, heated, and carbon-bearing gas is infiltrated through the preform. The preform

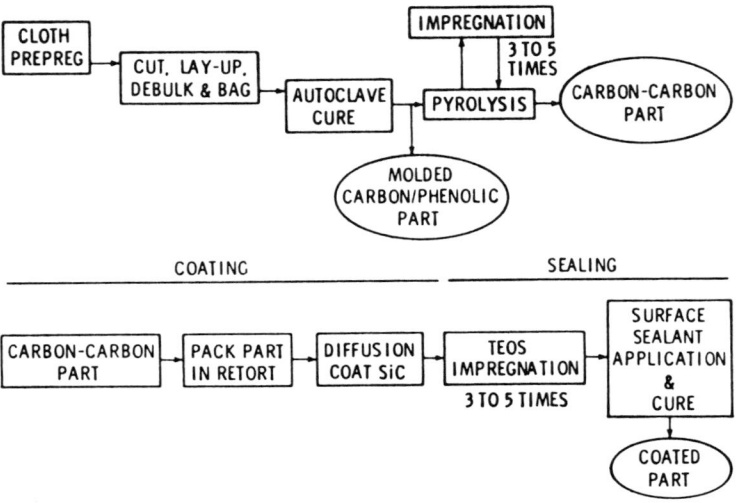

Figure 14.4 Fabrication steps involved in the manufacture of a two-dimensional carbon-carbon part (TEOS = tetraethylorthosilicate).

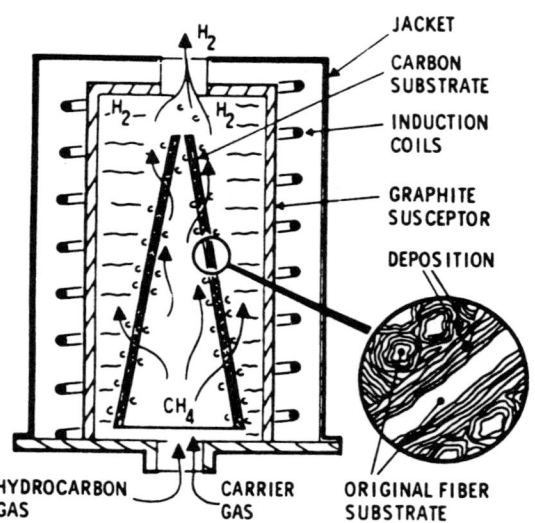

Figure 14.5 Isothermal chemical vapor deposition to infiltrate a fibrous carbon substrate.

CARBON-CARBON COMPOSITES

is at a suitable temperature and pressure, approximately 1100 °C (2012 °F) and 50 torr pressure to crack the gas, usually methane and hydrogen. Upon cracking, carbon deposits on the filaments. The process is indicated in Figure 14.5 for a conical preform. There are many variations of these processes. The part is usually graphitized at a higher temperature after the CVD processing.

PROPERTIES

The comparison of the mechanical properties of such carbon-carbon composites with the properties of metals and ceramics in Figure 14.6 clearly shows the advantages of carbon-carbon at high temperatures. Carbon-carbon composites are less susceptible than ceramics to thermal or mechanical shock failure. However, they are mechanically soft, cannot tolerate abrasive forces or handling, and must be protected from oxidizing environments. The materials are also expensive and can only be employed on high-value assets. Successful and economic applications are brake pads for military and commercial aircraft and nozzle liners for rocket motors. Aircraft afterburner nozzles are also made from carbon-carbon composites.

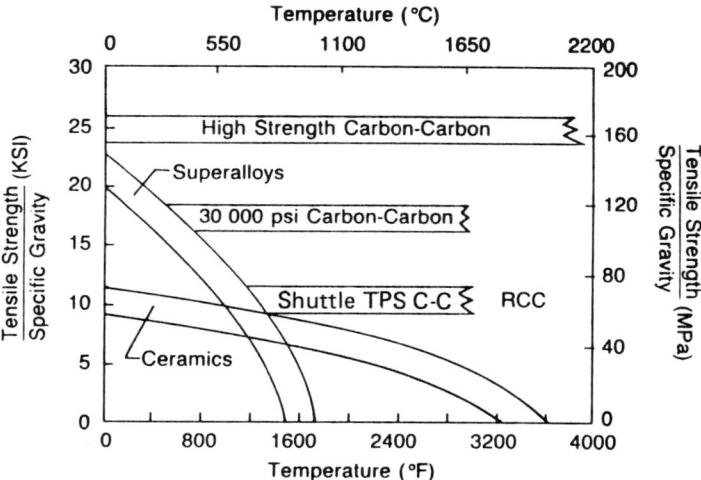

Figure 14.6 Strength-to-density ratio for several classes of high-temperature materials.

References

1. P. Rothenberg, *Complete Book of Ceramic Art*, Crown Publishers, Inc., New York, NY, 1972.
2. W.D. Kingery and B. Vandiver, *Ceramic Masterpieces*, The Free Press, New York, NY, 1986.
3. ASEA, ASEA Pamphlet AQ 20-104 E, 2nd ed., Metallurgical Division, Vasteras, Sweden.
4. W.D. Kingery, H.K. Bowen, and D.R. Uhlmann, *Introduction to Ceramics,* 2nd ed., John Wiley and Sons, New York, 1976.
5. M.E. Woods, W.F. Mandler, Jr., and T.L. Scofield, "Designing ceramic insulated components for the adiabatic engine," *Cer. Bull.* *64*, 2. Feb. 1985, American Ceramic Society, Columbus, OH.
6. M.E. Woods and T.L. Scofield, "Designing adiabatic engine components," *Proc. International Symposium on Ceramic Components for Engine*, KTK Publishers, Japan, 1983.
7. Garrett Turbine Engine Co., Ford Motor Company, AGT 101 report, Doc. No. MS7169, NASA Contract DEN3-1 67.
8. C. Kittell, *Introduction to Solid State Physics*, John Wiley and Sons, 3rd ed., 1966.
9. J.M. Herbert, *Ferroelectric Transducers and Sensors,* Gordon and Breach Science Publishers, New York, NY, 1982.

REFERENCES

10. R.C. Buchanan, *Jrnl. Am. Cer. Society, 42, 12,* p. 644, 1959.
11. W. Richerson, *Modern Ceramic Engineering,* Marcel Dekker, Inc., New York, 1982.
12. W.D. Kingery, *Ceramic Fabrication Processes*, MIT Press, Cambridge, MA, 1963.
13. R.L. Taylor and R.N. Donadio, "Vapor deposited IR materials," *Laser Focus, 17* (7), 1981.
14. R.N. Donadio, J.F. Connolly, and R.L. Taylor, "New advances in CVD IR transmitting materials," *Proc. SPIE, 297,* SPIE, Bellingham, WA, 1981.
15. G.E. Dieter, Jr., *Mechanical Metallurgy*, McGraw Hill Book Co., New York, NY, 1961.
16. L.R. Swank, et al., "Ceramic life prediction methodology final report," DOE, Contract No. DAAG 46-77-C-0028, Report No. MTL TR86-16, May, 1986.
17. "Design methodology for ceramic vanes and blades," GE Co. CRD TIS Report No. 78CRD016.
18. J.T. Neil, et al., "Fabrication of turbine components and properties of sintered silicon," Report No. 82-GT-252, ASME, New York, NY.
19. S.W. Freiman, "Brittle fracture behavior of ceramics," *Cer. Bull. 67,* 2, 1988. Am. Cer. Society, Columbus, Ohio.
20. L.M. Sheppard, "Ceramic engines are hot," *Mat. Eng.,* Oct. 1984, Penton Publishing, Cleveland, OH.
21. Kyocera Corporation, "Mechanical and industrial ceramics," CATD3/2T8710THA/3004E, Kyoto, Japan, 1987.
22. P.C. Smith, "Making ceramics tougher, *Mech. Eng.,* Jan. 1987, Penton Publishing, Cleveland, OH.
23. G.R. Anstis, et al., "A critical evaluation of indentation techniques for measuring fracture toughness," *Jnl. Amer. Cer. Soc. 64, 59,* Sept. 1981.
24. A.G. Evans and E.A. Charles, "Fracture toughness determinations by indentation," *Jnl. Amer. Cer. Soc. 59, 7-8,* July-August, 1976.
25. D. Munz, R.T. Busby, and J.L. Shannon, "Fracture toughness calculated from maximum load in four point bend tests of chevron notch specimens," *Intnl. Jnl. of Fracture,* Feb. 1980.
26. D. Munz, R.T. Busby, and J.L. Shannon, "Performance of chevron-notch short bar specimen in determining fracture toughness of Si_3N_4 and Al_2O_3," *ASTM Jnl. Testing and Eval.,* June 1979.
27. L.H. Van Vlack, *Elements of Materials Science*, 2nd ed., Addison-Wesley Publishing Co., Reading, MA, 1964.
28. S.W. Freeman, "Brittle behavior of ceramics," *Cer. Bull. 67, 2,* 1988.

REFERENCES

29. G. Greskovich and S. Prochazka, "A review of important observations on the sintering of SiC and Si_3N_4 ceramics," TIS 86CRD180, GE Co., CRD, Dec. 1986.
30. R. Komanduri, "Cutting tool materials," TIS 82CRD176., GE Co., June 1982.
31. C.E. Lewis, "Ceramics fire the imagination," *Mat. Eng.*, Penton Publishing, Cleveland, OH 1986.
32. V. Lanteri, A.H. Heuer, and T. E. Mitchell, "Tetragonal phase in system ZrO_2-Y_2O_3," *Am. Cer. Soc.*, 1984.
33. N. Claussen, M. Ruhle, and A.H. Heuer, editors, "Science and technology of zirconia II," *Am. Cer. Soc.*, 1984.
34. A.H. Heuer and M. Ruhle, "Phase transformation in ZrO_2 containing ceramics," *Am. Cer. Soc.,* 1984.
35. S. Musikant, H. Rauch, and E. Feingold, "Transformation toughening of ceramics for engines," Proceedings of the 21st Automotive Technology Development Contractors' Meeting, Report P-138, *Soc. of Auto. Engrs.*, Warrendale, PA, Nov. 1983.
36. J. G. Baldoni and T. Buljan, "Ceramics for machining," *Cer. Bull., 67, 2,* Feb. 1988.
37. R.E. Loehman and A.P. Tomsia, "Joining of ceramics," *Cer. Bull., 67, 2,* Feb. 1988.
38. L.M. Sheppard, "Microanalysis: solving microstructural mysteries," *Cer. Bull. 68, 6,* 1989.
39. "Diamond inserts you can afford," *Tooling and Production,* Feb. 1974.
40. R.P. Banas, E.R. Gzowski, and W.T. Larsen, "Processing aspects of the space shuttle orbiter, ceramic reusable surface insultion," 7th Annual Conference on Composites and Advanced Materials, *Am. Cer. Soc.,* Jan. 1983.
41. Lockheed Missile and Space Co., "Space shuttle high-temperature reusable surface insulation (HRSI) Fact Sheet," Nov. 1982.
42. C.E. Lewis, "Conductive ceramics," *Mat. Eng.* June, 1987.
43. G. Fisher, M. Schober, "Superconductor research pace quickens," *Cer. Bull. 66, 7,* 1987.
44. G. Fisher, "Superconductor mysteries unravel as developments proceed," *Cer. Bull. 67, 4,* 1988.
45. G. Fisher, "New technologies bolster electronic ceramics economics, *Cer. Bull. 63, 4,* April 1984.
46. Greenleaf Corporation, Advertisement, *Machine and Tool Blue Book,* Jan. 1986.

47. B. Schwartz, "Microelectronics packaging: II," *Cer. Bull. 63, 4,* April, 1984.
48. J.R. Kidwell, D.M. Kreiner, "AGT101 Advanced gas turbine technology development project," Proceedings 22nd Automotive Technology Development Contractors' Coordination Meeting, Report P155, *Soc. Auto. Eng.,* Oct. 1984.
49. D.W. Carruthers, et al., "Fabrication of silicon nitride turbine engine components," Proceedings 21st Automotive Technology Development Contractors' Coordination Meeting, Report P-138, *Soc. Auto. Eng.,* Nov. 1983.
50. L.C. Lindgren and D.A. Turner, "Ceramic component design," Proceedings 23rd Automotive Technology Dev. Contr. Coord. Mtg., Report P-165, SAE, Inc., Oct. 1983.
51. R. Ohnsorg, et al., "Ceramic component fabrication," Proceedings 23rd Automotive Technology Dev. Contr. Coord. Mtg., Report P-165, SAE, Inc., Oct. 1983.
52. "Preform architecture for ceramic-matrix composites," *Cer. Bull. 68, 2,* 1989.
53. J.D. Buckley, "Carbon-carbon, an overview," *Cer. Bull. 67, 2,* 1988.
54. T. Baumeister, ed., *Marks' Mechanical Engineers' Handbook,* McGraw Hill Book Co., New York, New York, 1988.
55. "Materials selector," *Mat. Eng.,* Penton Pub. Co., Cleveland, OH, 1987.
56. C.O. Smith, *Nuclear Reactor Materials*, Addison-Wesley Publishing Co., Reading, MA 1967.
57. W.J. Lackey, et al., "Ceramic coatings for heat engine materials-status and future needs." Proceedings of the 1st International Symposium on Ceramic Components for Engine, KTK Scientific Publishers, Japan, 1983.
58. Tai-il Mah, et al., "Recent developments in fiber-reinforced high temperature ceramic composites," *Cer. Bull., 66, 2,* Feb. 1987.
59. D.L. Hartsock and A.F. McLean, "What the designer with ceramics needs," *Cer. Bull. 63, 2* Feb. 1984.

Index

AA750 adiabatic engine, 89
Abrasives, 15
Additives, 58
Adiabatic diesel engine, 35, 89, 179
Advanced gas turbine (AGT) project, 35
  AGT-5 ceramic gas turbine engine, 165
  AGT 100, 176
  AGT 101, 42, 167
Alumina, 12
  coated cemented carbide, 141
  composites, 137
Arch, 10
Attrition mill, 57
Automotive ceramics, 165

Ball mill, 56
Barium-yttrium-cooper oxide, 153
Bauxite, 54
Bednorz (Nobel prize winner), 153
Binder, 54
Boehmite, 54
Bone ash, 16
Borazon, 137
Brazes for bonding of ceramics, 125
Brick, 17
Brittleness, 79

Capacitor dielectrics, 164
Carbon, 187
  preforms, 189

INDEX

Carbon-carbon,
  airfoils, 185
  composites, 185
  process (chart), 192
  properties, 193
CASE, 92
Catalytic converters, 165
Ceramic automotive Stirling engine, 92, 98
Ceramic coated cutting tools, 144
Ceramic hardware cycle (chart), 173
Ceramic history (chart), 4
Ceramic turbocharger, 182
Ceramics,
  applications, 32, 98
  definition, 1
  traditional versus fine, 33
Chemical machining, 77
Chemical vapor deposition, 71
Chevron notch tests for fracture toughness, 106
Chu, P.C.W., 153
Clay, 15
Computed tomography (CT), 132
Concrete, 17
Conductors, ceramic, 163
Cordierite, 165
Covalent bonded, fracture toughness, 109
Crack propagation, impeding, 112
Critical stress, 100
Critical current density, 155
Cubic boron nitride, 137
Curie temperature, 43
Cutting bit, 6

Cutting tools, 33
  ceramic, 51, 137, 140
CVD, 71
  schematic, 75

Densification, 55, 59
Design methodology, of ceramic engine components, 174
Diamond
  cutting tools, 144
  hardness, 109
Dielectric, 5
  ceramic, 163
Drain casting (schematic), 64
Ductile-brittle transition, 80
Dye penetrant, 133

Earthenware, 1
Electrical discharge machining, 77
Electronic ceramics, 42, 161
Eutectic, 61
Extruder (schematic), 61

Fabrics, for reinforcement, 191
Feldspar, 15
Ferrites, 46
Ferroelectrics, 42
Fibers
  for ceramic-matrix composites, 190
  for reinforcement of ceramics, 117
Films
  thick, 161
  thin, 161
Fine ceramics, 9
Flaw size, 100
Flaw sensitivity, 79
Flint, 15

INDEX

Fluid energy mill, 57
Flying buttress, 10
Fracture, 6
Fracture surface, 112
Fracture toughness, 11, 101
FRCI-12, 149
Furnace, 9
Furnace, glass, 15

GaAs, fracture toughness, 109
Garnet, 46
Glass, 2
Glass-ceramic, 3
Glassy bond, 126
Glaze, 18
Granite, 10
Graphite, 187
Green body, 58
Griffith relationship, 99

Heat engines, 35
   applications, 88
Hot isostatic pressing (HIP), 71
Hot pressing (HP), 71
HRSI, 149
Hysteresis of ferroelectrics, 43

Image enhancement, 130
Indentation test, 101
Injection molding process (flow chart), 63

Japan, 31
Joining of ceramics, 123
Josephson junction, 156

La_2CuO, 153
Lapping, 77
LI-2200, 149

Limestone, 17
Liquid-phase sintering, 61
Lithium aluminosilicate (LAS), 81
LRSI, 149

Machining of ceramics, 74
Magnetic resonance imaging (MRI), 134
Metallurgy, 10
Microfocus X-ray, 130
Microstructure, 12
Moderator, 50
Muller (Nobel prize winner), 153
Mullite, 12, 54

Net shape, 59
Nextel, 151
Niobium-alpha alumina bonding, 125
Nicalon, 117
Nondestructive testing (NDT), 129
Nondestructive evaluation (NDE), 132
Nonoxide ceramics, joining, 127
Notre Dame Cathedral, 10
Nuclear reactor applications, 50

Oxidation inhibition of carbon-carbon composites, 191

Particle size, 56
Phase equilibrium diagram, alumina/silica, 54
Phase transformation toughening, 119
Photoetching, 77

Piezoelectric, 44
  oscillator, 45
  tuning fork, 45
Plasma spray, 71
Poling, 43
Polycrystalline ceramics, 43
Porcelain, 1, 18
Porosity, 56
Portland cement, 16
Potter's wheel, 53
Powders
  ceramic, 55
  formation, 58
Pressing (flow chart), 66
Proof testing, 82
Properties of ceramics, 30
Pyroceram, 3
Pyroelectric effect, 44

Quartz, 12

Radiography, 129
Reactive metal, 125
Redox reaction, 124
Reflector (in nuclear reactor), 50
Refractory, 1, 56
Resistors, ceramic, 163
Rochelle salt, 45
Rock wool, 5
Rocket motor nozzles, carbon-carbon, 187
Rocks, 10

Sandblasting, 77
Shielding, nuclear, 50
Shrinkage, 59
Si_3N_4, 55, 56
  crack propagation in, 114
SiAlON (sialon), 141
Silica, 17

Silicon carbide (SiC), 2, 85
Silicon nitride (*see also* Si_3N_4), 137
Single crystal, toughness, 107
Sintering, 59
  aid, 54
  mechanisms (chart), 64
Sodium vapor lamp, enclosure, 56
Solders for bonding of ceramics, 125
Space Shuttle
  insulation tiles, 147
  window strength, 84
Spinel, 46
Startup stresses, in engine, 167
Statistical analysis, failure, 81
Stones, 10
Stoneware, 1
Strain isolator, Space Shuttle tiles, 151
Stress analysis, 85
Stress intensification factor, 100
Structural clay, 15
Superconductors, 153
  applications, 156
  fabrication, 157
Surface acoustic wave (SAW) delay line, 45
Surface analysis, 135
Surface flaws, 122
Surface treatments for strengthening, 122

Tape-forming (schematic), 65
Terra-cotta, 1
Thermal analysis, 85
Thermal shock, 6
Tile, 18
Tomography, X-ray computed, 132

INDEX

Tool wear, 137
Toughening of ceramics, 111
Tourmaline, 45
Transformation toughening, schematic, 121
Transformation-toughened ZrO_2, 52
Transition metals, 46
Transmission electron microscopy, 135
Tunneling, 156
Turbine wheel, 6
Turbine scroll assembly (figure), 181

Valve stem guide, 181
Valve seat, 181
Varistor, 47

Vibratory mill, 57
VLSI, 163

Wedgewood, 5
Weibull modulus, 81
Weibull distribution, 81
Whisker reinforced Al_2O_3, 52
Whisker reinforcement, 114
Whiteware, 15

Yarns for reinforcement, 191
$YBa_2Cu_3O_7$, 155
YIG, 46

Zirconia, 119
ZnSe, fracture toughness, 109
ZrO_2-Y_2O_3 phase diagram, 119, 120
ZrO_2, 119

About the Author

Solomon Musikant is President, TransCon Technologies, Inc., Paoli, Pennsylvania. He is also Adjunct Professor in the Graduate Engineering School at Villanova University, Villanova, Pennsylvania. Previously, Dr. Musikant served as Program Manager for Advanced Materials, General Electric Co., Astro Space Division, where he led research and development of advanced materials for spacecraft (1968-1990). The author or editor of over 60 publications, including *Optical Materials: An Introduction to Selection and Application* and *Optical Materials: A Series of Advances, Volume 1* (both titles, Marcel Dekker, Inc.), Dr. Musikant holds six U.S. patents for glass and ceramic processes and applications. A Fellow and Director of SPIE, The International Society for Optical Engineering, he received the B.S. degree in mechanical engineering from the Cooper Union Institute of Technology, New York, New York, M.S. degree in metallurgy from the Stevens Institute of Technology, Hoboken, New Jersey, and Ph.D. degree (1967) in materials science from Lehigh University, Bethlehem, Pennsylvania.